T0135057

Smart Innovation, Systems and Technologies

Volume 95

Series editors

Robert James Howlett, Bournemouth University and KES International,
Shoreham-by-sea, UK
e-mail: rjhowlett@kesinternational.org

Lakhmi C. Jain, University of Technology Sydney, Broadway, Australia;
University of Canberra, Canberra, Australia; KES International, UK
e-mail: jainlakhmi@gmail.com; jainlc2002@yahoo.co.uk

The Smart Innovation, Systems and Technologies book series encompasses the topics of knowledge, intelligence, innovation and sustainability. The aim of the series is to make available a platform for the publication of books on all aspects of single and multi-disciplinary research on these themes in order to make the latest results available in a readily-accessible form. Volumes on interdisciplinary research combining two or more of these areas is particularly sought.

The series covers systems and paradigms that employ knowledge and intelligence in a broad sense. Its scope is systems having embedded knowledge and intelligence, which may be applied to the solution of world problems in industry, the environment and the community. It also focusses on the knowledge-transfer methodologies and innovation strategies employed to make this happen effectively. The combination of intelligent systems tools and a broad range of applications introduces a need for a synergy of disciplines from science, technology, business and the humanities. The series will include conference proceedings, edited collections, monographs, handbooks, reference books, and other relevant types of book in areas of science and technology where smart systems and technologies can offer innovative solutions.

High quality content is an essential feature for all book proposals accepted for the series. It is expected that editors of all accepted volumes will ensure that contributions are subjected to an appropriate level of reviewing process and adhere to KES quality principles.

More information about this series at http://www.springer.com/series/8767

Hendrik Knoche · Elvira Popescu
Antonio Cartelli

Editors

The Interplay of Data, Technology, Place and People for Smart Learning

Proceedings of the 3rd International
Conference on Smart Learning
Ecosystems and Regional Development

 Springer

Editors
Hendrik Knoche
Department of Architecture, Design,
  and Media Technology
Aalborg University
Aalborg
Denmark

Antonio Cartelli
Viale dell'Università - Rettorato -
  (Campus Universitario) Loc.
Università degli studi di Cassino e del Lazio
  Meridionale
Folcara
Italy

Elvira Popescu
Computers and Information Technology
  Department
University of Craiova
Craiova, Dolj
Romania

ISSN 2190-3018   ISSN 2190-3026   (electronic)
Smart Innovation, Systems and Technologies
ISBN 978-3-030-06350-4   ISBN 978-3-319-92022-1   (eBook)
https://doi.org/10.1007/978-3-319-92022-1

Printed on acid-free paper

This Springer imprint is published by the registered company Springer International Publishing AG part of Springer Nature
The registered company address is: Gewerbestrasse 11, 6330 Cham, Switzerland

Preface

This volume contains the papers presented at SLERD 2018: 3rd International Conference on Smart Learning Ecosystems and Regional Development.

Following successful first (Timisoara, Romania in 2016) and second editions (Aveiro, Portugal in 2017), SLERD 2018 was hosted by the Interaction Lab from the Department of Architecture, Design and Media Technology at Aalborg University, Aalborg, during May 23–25, 2018.

The Interaction Lab is an interdisciplinary research group focusing on innovation in the design of new interaction approaches for human-centered digital media applications. The laboratory specializes in applied research with relevant regional, national, and international stakeholders in the areas of health, smart learning, and robotics and has build up a strong research expertise in interaction with special needs groups like citizens with traumatic or congenital brain damage, dementia, dyslexia, as well as indigenous groups, and kindergarten and school children.

The conference was co-organized by the Association for Smart Learning Ecosystems and Regional Development (ASLERD), an international non-profit interdisciplinary, democratic, scientific-professional association that is committed to support learning ecosystems to get smarter and play a central role to regional development and social innovation. "Smart," thus, are not simply technology-enhanced learning ecosystems but, rather, learning ecosystems that promote the multidimensional well-being of all players of learning process (i.e., students, professors, administrative personnel and technicians, territorial stakeholders, and, for the schools, parents) and that contribute to the increase of the social capital of a "region," also thanks to the mediation of the technologies. ASLERD, thus, aims at generating a concrete impact by understanding learning ecosystems and accompanying design for "smartness," fostering the development of policies and action plans, supporting technological impact well beyond prototypes and pilots, promoting networking and opportunities to discuss and debate like the SLERD yearly conference.

SLERD 2018 aimed at promoting reflection and discussion concerning R&D work, policies, case studies, entrepreneur experiences with a special focus on understanding how relevant the smart learning ecosystems (schools, campus, working places, informal learning contexts, etc.) are for regional development and social innovation and how the effectiveness of the relation of citizens and smart ecosystems can be boosted. The conference had a special interest in understanding how technology-mediated instruments can foster citizen engagement with learning ecosystems and territories, namely by understanding innovative human-centric design and development models/techniques, education/training practices, informal social learning, innovative citizen-driven policies, technology-mediated experiences and their impact. This set of concerns contributes to foster the social innovation sectors and ICT and economic development and deployment strategies alongside new policies for smarter proactive citizens.

The selected scientific papers aim to understand, conceive, and promote innovative human-centric design and development methods, education/training practices, informal social learning, and citizen involvement. These grouped around three main conference themes: (i) places for learning, (ii) learning technologies, and (iii) human-centered design.

SLERD 2018 contributes to foster the social innovation sectors, identifying and discussing ICT and economic development and deployment strategies alongside with new policies for smarter proactive citizens. The proceedings are relevant for researchers and policy makers alike.

In summary, SLERD 2018 offered an exciting program that provided an excellent overview of the state of the art in smart learning ecosystems and was an occasion for bringing research forward and creating new networks.

We are very happy about the final selection of papers, which would not have been possible without the effort and support of our excellent Conference and Program Committees, including more than 50 international researchers. We would like to thank all the ones who, in different roles, have contributed their time to organize the event with enthusiasm and commitment.

April 2018   Hendrik Knoche
                                                                                       Elvira Popescu
                                                                                     Antonio Cartelli

Organization

Program Committee

Joao Batista	University of Aveiro
John Carroll	The Pennsylvania State University
Antonio Cartelli	Università degli Studi di Cassino e del Lazio meridionale
Mihai Dascalu	University Politehnica of Bucharest
Ines Di Loreto	UTT—Université de Technologie de Troyes
Monica Divitini	Norwegian University of Science and Technology
Gabriella Dodero	ASLERD
Carlo Giovannella	Sapienza University of Rome
Ralf Klamma	RWTH Aachen University
Hendrik Knoche	Aalborg University
Stefania Manca	ITD-CNR
Antonio Moreira Teixeira	Universidade de Lisboa
Antonella Nuzzaci	University of L'Aquila
Donatella Persico	ITD-CNR
Elvira Popescu	University of Craiova
Fernando Ramos	University of Aveiro
Matthias Rehm	Aalborg University
Annika Wolff	Lappeenranta University of Technology
Imran Zualkernan	American University of Sharjah

Additional Reviewers

Pasov, Iulia
Ravicchio, Fabrizio
Toma, Irina

Contents

Smart Learning Ecosystems

Improving Massive Alternance Scheme: The Paradigmatic Case History of the *Incubator of Projectuality* at the Ferrari School of Rome

Carlo Giovannella[1,2(✉)], Ida Crea[3], Giuseppe Brandinelli[3], Bianca Ielpo[3], and Cristina Solenghi[3]

[1] ISIM_Garage, Department of History, Cultural Heritage, Education and Society, University of Rome Tor Vergata, Rome, Italy
carlo.giovannella@uniroma2.it
[2] ASLERD, Association for Smart Learning Ecosystems and Regional Development, Rome, Italy
[3] IIS E. Ferrari, Rome, Italy

Abstract. Since three years the Italian Ministry of Education (MIUR) has introduced a massive alternance scheme, completely new to Europe, whose implementation still shows several critical issues and requires a special effort by all potential players - schools, associations, enterprises, local communities, etc. - to design and experiment models and strategies capable to mitigate them. In this article we present the first validation of an approach, based on the simulation of innovation processes, that can be considered innovative in the context of the Italian School Work Alternance (SWA) scheme. The outcomes of the experimentation are encouraging and show an average increase of more than one point over ten in the student satisfaction with respect to that of peers who have experienced other schemes of SWA activities during the previous school year. Not by chance our proposal has been selected as best practice by the local association of the entrepreneurs. Despite of such positive result, as discussed in the body of the paper, there exists a considerable room for improvements.

Keywords: Massive alternance scheme · Innovation process
Smart learning ecosystems · Schools' smartness

1 Introduction to Societal Context and Problems

A sustainable transition from school to work is one of the priorities of the European 2020 strategy [1, 2] and learning ecosystems are expected to support and guide the students in the development of adeguate abilities to face up with the transformation of the working chains (i.e. I4.0) and with societal challenges. Educators, technicians, parents and territorial stakeholders are all expected to sustain and contribute to the increase of students' long standing competences and, thus, to their level of employability in a world where the productive processes are renewing at an incredible speed. Not by chance, central to the European strategy is the *work based learning* that has been implemented differently from country to country [3]. Well known is the dual educational

© Springer International Publishing AG, part of Springer Nature 2019
H. Knoche et al. (Eds.): SLERD 2018, SIST 95, pp. 3–14, 2019.
https://doi.org/10.1007/978-3-319-92022-1_1

system adopted in Germany and Austria, that combines well regulated apprenticeships in a company with education at a vocational school [4, 5]. Other European countries have adopted "lighter" alternance schemes (for the definition of the term *alternance* see [6]) where the duration of work based learning may range from weeks to one year and students, usually, have not an apprentice status. Benefits and obstacles of the work based learning have been well documented by European Training Foundation [7]. Possible benefits span from skills and competences progressing to the development of a professional identity, to greater employment opportunities and to productivity gains. On the other hand beside benefits also relevant challenges have to be faced: (a) the development of an efficient placement that requires the prioritization of sectors and qualifications; (b) the identification of engaging opportunities; (c) an adequate quality of the pre-training, that may not be easy to guarantee when founds are limited; (d) the involvement of micro and small enterprises that may not see sufficient pay-backs to get engaged in alternance schemes.

Despite the above well-known criticalities, in 2014 the Italian government decided to tackle a huge challenge: the implementation of a massive alternance scheme – School-Work Alternance (SWA) - involving all students of the last three years of both high and vocational schools: 200 h for high school students and 400 h for vocational school students. The law prescriptions imply a huge effort from either schools and productive system that, actually, were both not well equipped to withstand the impact of the law 107, better know as "Buona scuola" (good school) law [8]. In fact, considering that in Italy about 0,7 ML of students are attending the last three years of the high school cycle and about 0,8 ML of students the last three years of the vocational schools, the expected work-based learning days foreseen by SWA are overall about 62,5 ML that in principle should be offered by about 5.7 ML enterprises and productive activities, most of which (more than 90%) are micro and family enterprises [9]. This is simply out of any practical feasibility and the crisis of the system should had not been very difficult to predict. Under these circumstances it is not strange that to cover the needs of the massive alternance scheme also no profit associations, public administrations and universities have been involved in SWA, with questionable results, also not difficult to predict. Not by chance, a survey conducted during school year 2016–2017 on a sample of schools of the South-Est Rome area made emerge a set of criticalities that mirrors a more general situation confirmed by a huge amount of grey literature, journalist inquires and, overall, by a report that has been published in 2016 by Unindustria [10]. To stress the seriousness of the problem, that unavoidably affects the perception of the schools' smartness, it is worthwhile to point out that the indicator related to the student satisfaction about SWA came out to be the lowest one in all schools involved in the survey. The criticalities that emerged from the survey mapped quite well on the challenges described in [7] that can be resumed as follow: (a) unstructured governance; (b) timing; (c) unchecked preconditions and prerequisites; (d) limited significance of the activities and unsatisfactory rewards.

Due to the unstructured governance and to the limited significance of most of the SWA experiences, at present, students and entrepreneurs tend both to perceive SWA as a burdening interference of the daily activities. Unchecked preconditions (e.g. reciprocal acquaintance, undefined activity plan and goals, etc.) and prerequisites (e.g. tutors training and evaluation of student competences) are among the main complaints that emerged from the survey. Unfortunately also the opinion of a large part of the parents on SWA activities is quite critical and indicates the existence of a strong gap among educational agencies (the detailed analysis of parents' opinions is out of the scope of this paper and will be presented in a paper to come).

Recently, since three years from the promulgation of the "Buona Scuola" law, MIUR seems finally to get conscious about the effects induced by the adoption of the massive alternance scheme and, because of this, is trying to redefine its focus and role, now "reduced" to an "innovative learning methodology" the goals of which, however, remain still not clearly defined. In the while, waiting for future guidelines, the general tendency is to use the SWA experience as a period of orientation to inform students about entrepreneurship, safety at work and, as well, about the opportunities offered by universities' curricula and research activities. Such interpretation of the SWA activities, however, are far away from the expectations of students and entrepreneurs that emerged from both our survey and the Unindustria's report. Students expect to:

- develop LIFE skills, possibly certified ones;
- experience personalized SWA activities, capable to develop also excellence;
- get adequately challenged by practical activities, also to test competences and skills that they believe to have developed, at least partially.

Entrepreneurs, on the other hand, wish to

- get involved in the design of SWA experiences;
- get involved in the certification of students;
- and, possibly, get a payback in terms of competitive advantages and/or detaxation.

Putting all together the key issues described above, we can formulate a *great challenge* (rather than a research question): *the elaboration of win-win strategies aimed, as much as possible, at mitigating criticalities and satisfying expectations of all players involved in SWA activities*. Of course, succeeding in this great challenge, in turn, will help the schools to increase their level of smartness. In ASLERD context [13], in fact, smartness of learning ecosystems does not simply means "technology enhanced". The smartness is a more complex multilayered construct related to the "wellbeing" of all players operating in the ecosystems, that are also expected to be in relation with the territory. Smartness, thus, is affected by the improvement of any relevant aspects of the learning processes and ecosystem functioning, especially if connected with territorial development and social innovation.

In the next paragraph, thus, we describe our proposal based on the realization of *incubators of projectuality (IoP)*. The validation of the proposal is described in Sect. 3, while the lesson learnt and future perspectives are presented in Sect. 4.

2 The Incubator of Projectuality

The *incubator of projectuality* is intended to involve students in the *simulations of innovation processes* with the aim to prepare them for future internships and/or enterprise simulations. In fact innovation processes, being also *design based*, allow the students to acquire in a controlled manner a sufficient level of *design literacy* and as, a consequence, also a significant set of *horizontal competences* (LIFE skills) the acquisition of which, as well known, is stimulated by activities and methodologies typical of a design process, see the mapping reported in ref. [11]. Design processes could be also largely personalized and contextualized to satisfy needs and expectations of all players involved in SWA, and by nature, when focused on innovation, are challenging; moreover, if supported by on-line collaborative learning-working environments allow students to develop practical and in-depth digital skills.

Innovation processes can be developed for a large part in situ (i.e. within the schools), without a heavy involvement of company resources. Companies on the other hand can contribute to their co-design, to the certification of competences and to increase the training significance through the involvement of testimonials and the narration of the best company practices. Due to the use of on-line collaborative learning-working environments, companies have also the opportunity to supervise the processes and pre-screen students that could be possibly involved in future internships.

It is worthwhile to underline that although the introduction of design processes into the schools is not new, the mastery of an adequate design literacy among Italian school teachers is not common (except for the schools focused on specific design domains). In fact the word "design" is usually associated by teachers to the "learning design" and not to what should be considered a cornerstone of the educational path of each student. In the past there have been an attempt to introduce into Italian schools the basics of the design literacy by means of the Service Design Thinking approach [14] - that as well know is only one of the possible declination of the Design approach, focused on services [15]. The outcomes, however, have been quite poor and unfortunately have generated the propagation of a certain amount of misconceptions as can be easily verified by browsing the grey literature and reports available on the web [16]. The *design approach to the life* is a capability that requires a long practice to be achieved and mastered before anyone could teach it.

In other cases, to introduce design basics into the schools, people has adopted a framework of reference nowadays very popular: the "design thinking" [17, 18]. However one has to be aware that "design thinking" has not added anything new with respect to the idea that the *design* is central to any human activity that intends to modify the world [21]: before the introduction of the "design thinking" one would had made reference to the *centrality of the project*. This latter, in fact, is central to understand, innovate and manage any situation and context - whatever their degree of complexity - and does not start from the divergent problem solving but, rather, from exploration and problem setting, especially if the contexts are complex and the problem are ill-posed or wicked ones [22, 23]. A third element that is usually included in the "design thinking" is the centrality of the people that, actually, has been fully developed in the *design for the experience* context and that when integrated with the *theory of complexity* makes emerge the richness of the *co-evolution:* people and places in continuous osmotic interaction (i.e. *people in place centered design*).

Apart from this quick overview on the past attempts to introduce the design culture into schools - useful to better understand the framework of reference - it is important to stress that at the time we started the *IoP* process at the IIS Ferrari - school year 2016–17 - "design thinking" or other design frameworks were not yet applied to the SWA context. Nowadays one can find on the web several evidences (mainly announcements or short news) about the use of "design thinking" in SWA activities, but as far as we know no one has validated them against other activities as we have done (see next paragraph). Moreover, no one has provided a mapping between activities carried on during the design activities and the potential acquisition and certification of competences, as we have done in ref. [19].

Another aspect that marks the difference between the experience we developed at the IIS Ferrari and "design thinking" based activities is the adoption of a very flexible design process capable to keep continuously engaged the students: the Organic Process, OP, that is in use at the University of Tor Vergata since more than 10 years in the course of Multimodal Interfaces and Systems that implements innovation processes based on the *design for experience* approach [20]. The detailed description of OP is out of the scope of this paper and can be found in ref. [12]. To mark the difference between our choice and the design thinking approach it is enough to remind here that OP is organized into "layers of functionalities" and has been inspired by "living organisms" that all, at almost any level of complexity, fulfill three basic functionalities: (a) investigate the environment to collect information & learn; (b) elaborate the information to design/ produce; (c) communicate the "products" of this elaboration by means of "behaviors" that, in the case of very complex organisms, can make use of highly structured languages and tools. These vital functions can be considered as collective activities that are expected to be always active during the entire time-window of the process' unfolding. Another important difference that has to be underlined concerns the testing and monitoring activities that in our case do not represent a phase but should be considered as integrated to and cutting across all three functional layers of OP. Testing and monitoring activities are crucial to the smooth progress of the process and are also always active.

The next step in the development of the IoP has been the definition of: phases/activities, deliverables, timing and the identification of the competences that are expected to be developed by the students. Phases and activities have been organized in three connected subprocesses, in principle one for each of the years along which the SWA scheme is expected to be developed:

- from the problem setting till the narration of the selected solution;
- from the development of the executive plan till the development of the prototype and the presentation of the demo;
- from the protection of the idea till its exploitation and the definition of plans for continuous innovation.

In this paper we will report on the first validation of our SWA solution that has been developed and tested during March–May 2017 with a group of 25 students attending the third year of the vocational schools in informatics of the IIS Ferrari of Rome. They have experienced the first subprocess listed above that included: design of data collection, data collection and analysis, problem setting and its narration, creative problem solving, solution selection and pitch presentation.

Before to start the SWA activities students have been introduced to the relevance of innovation. Immediately after they have ben asked to form groups of 5 students and to identify a target for their innovation process, among the following ones: (a) school innovation (internal target); (b) social innovation; (c) entrepreneurial activities (located in the surrounding territory); (d) other free choice targets.

All groups concentrated on the first two targets. Their choices demonstrate the potentialities of the IoP in supporting the increase of school's smartness and social innovation (see ASLERD vision [13]). Students that have chosen the first target took advantage of the participatory procedure that were carried on few months before to evaluate the school's smartness and made emerge criticalities/expectations from all players involved in the learning processes (participatory evaluation). The other groups had first to design a data collection process and then collect the data from which they took inspiration to develop the innovation process. The outcomes of the SWA experience were the five interesting concepts described in shorts here below:

- a reservation system - app based - to avoid queue at the school café, automate payments and support optimized food supplies;
- an app to support competence matching (supply-demand), evaluation and certification during SWA
- an app to support a gamified social parking
- a system to monitor the filling of garbage bins and optimize the rubbish collection
- an app to organize car sharing for disadvantaged categories like, for example, elderly people.

3   Evaluation and Outcomes

At the end of the experience we have distributed a questionnaire to collect the opinions of the students that took part in the development of SWA activities based on the simulation of innovation processes. It contains questions intended to collect quantitative evaluations of several factors (on a 1–10 scale) and, as well, open questions and request of comments intended to attach sense to the numerical evaluations. To be sure that the evaluation of the SWA experience was not driven by the performance of the tutors we have asked also to evaluate them. Unfortunately due to the limited number of pages available for this publication we had to drop the appendix containing the questionnaire. It will be published in a paper to come.

To work out the level of efficacy of our proposal in answering the grand challenge described in the previous paragraph, the outcomes of this questionnaire has been compared, see Table 1, with the outcomes of the two surveys that have been launched in February-March 2016 and in October–November 2017 to benchmark the level of smartness of the school, and that included also questions on SWA experiences. The surveys involved more than 80% of the student population (531 in 2016 and 527 in 2017). During the first survey we asked only one generic question on the SWA experiences to extract an indicator of satisfaction and make emerge possible criticalities while during the second one we have presented additional questions to evaluate also other aspects of SWA experiences (see Table 1): significance of the experience, school ability to manage SWA experience, overall level of challenge. These aspects have been evaluated also by the group of students that took part in the simulation of the innovation process (IoP).

Table 1. Mean values and standard deviations extracted from the distributions (scale 1–10) generated by the students involved in the surveys. IoP = incubators of projectuality. Columns 2016 (school year 2015–2016) and 2017 (school year 2016–2017) have been generated by students that already experienced SWA activities while 2016 P and 2017 P are generated by students that have not yet be involved in SWA and mirror, thus, students' perceptions and expectations about SWA. The *Challenge* indicator refers more in general to the challenges offered by the school to the students except for IoP. In this latter case it refers to the level of challenge offered by the simulation of innovation processes.

Index	IoP	2016	2017	2016 P	2017 P
Satisfaction	5,67±0,88	4,30±0,18	4,94±0,46	4,78±0,21	5,35±0,38
Significance	5,77±1,01	—	4,94±0,43	—	5,66±0,42
Governance	5,62±0,79	—	5,12±0,39	—	5,69±0,42
Challenge	5,42±1,09	4,71±0,27	5,38±0,39	5,20±0,41	6,05±0,31

Since the surveys we launched in February-March 2016 and in October-November 2017 involved the whole school population we had to make a distinction between the students that were not yet involved in SWA experiences and the others. In the case of the students that were not yet been involved in SWA experiences their inputs should be interpreted as perception and expectations about SWA.

To favor a better understanding of the mean values reported in Table 1, in Fig. 1 we have plotted the distributions from which the mean values of the parameter *Satisfaction* have been extracted. In 2016 the distribution of the student perception (red line) and truly *Satisfaction* (violet line) overlapped almost completely, apart for an increase of the very low values tail that is the cause of the slight decrease of the mean value of the indicator.

In 2017 the mean values of both students' expectations and satisfaction have consistently grown but in no cases the *Satisfaction* is higher than the one measured at the end of the IoP. This latter is greater than the one measured in 2016 by about 1,4 (statistically fully significative since larger than the sum of the standard deviations of the two compared distributions). The increase of 0,7 of the average satisfaction with respect to that emerged during the general survey of the 2017 - i.e. of the satisfaction of the students that participated in other SWA experiences - cannot be considered fully significative due to the large standard deviations but together with the increase of *significance* and *governance* indicators are strong clues in favor of an overall better quality of the approach based on the simulation of innovation processes.

Two aspects that deserve further comments are: (a) the higher mean value of perception/expectations with respect to the mean value of all indicators measured after SWA experiences and (b) the value of the *Challenge* indicator for year 2017 that is very similar to that measured after the IoP activity.

To get a better insight into the first aspect we have separated the distributions of the perceived level of *Satisfaction* generated by student cohorts of the first three years, see Fig. 2. The plots clearly indicate that students of the first year have strong expectations and that such expectations decrease during second and third years, with a mean value that $6,81 \pm 0,56$ decreases to $5,14 \pm 0,78$ and finally to $4,45 \pm 0,56$ during the beginning of the third year. A negative trend that can be considered statistically meaningful since the overall variation between the first and the third year is larger than 2,3. Similar trends have been observed also for the other indicators related to the SWA activities and demand for actions aimed at mitigating this negative tendency.

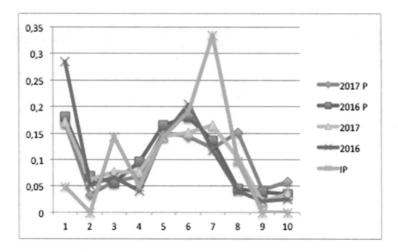

Fig. 1. Distributions of the Satisfaction about SWA experiences (2016 in violet, 2017 in light green and IoP in light blue) and the perception/expectations about SWA (2016 P in red and 2017 P in blue)

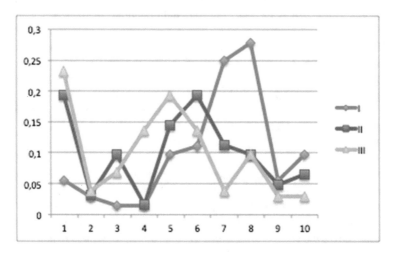

Fig. 2. Distributions of the perception/expectations about SWA for students attending the I, II and III years of IIS Ferrari.

Coming now to the value of the *Challenge* indicator that was very close to that measured after the IoP experience, an explanation can be found in the actions that have been taken by the school to sustain and valorize the excellences, following the outcomes of the 2016 survey: i.e. the event that has been organized at the end of the school year during which the best performances of the students have been recognized and rewarded. Possibly the same sense of gratification has been perceived by the students of the IoP. To this respect, it is interesting to note that the value of the *Satisfactory* indicator for IoP, as stated by the students, could have been increased in average by at least one more point if the questionnaires would have been delivered after the last meeting that involved representative of a software enterprise. This because she shown the enterprise's appreciation for the outcomes of the students' concepts, and this was very rewarding for the students.

To the students that attended IoP activities we have also asked (see appendix 1) "*if they were interested in a follow up of the experience and how much*". The answer was in average positive and the mean numerical score, on the scale 1–10, was $5.86 \pm 1,00$, slightly higher although not significantly higher that all the other indicators reported in Table 1. What were more interesting, however, were the motivations that emerged from comments, that we can condensate in the following two sentences: "*I'd like to continue because we conceived the concept*", "*I don't want to leave it as an incomplete project, I want to realize it*".

Overall most of the students considered the IoP's activities very positive because they had the feeling: (a) to take part in work oriented activities; (b) to develop concepts having a public utility (that in turn induced the desire to develop it further); (c) to learn useful methodologies and approaches (problem setting, brainstorming, collaborative work, pitch presentation, etc.)

As stated above we have checked also for a possible influence of the tutors' performance on the evaluation of the SWA activities based on IoP. It is interesting to note that the students' evaluation of the IoP experience is independent from the evaluation of the educators that were involved. In fact the mean values indicators related to the teaching, tutor and mentoring activities were in general much higher than those reported in Table 1: $6.39 \pm 0,91$ for *clarity in exposition*, $6.68 \pm 1,10$ for the *ability to highlight key issues*, $7.35 \pm 0,85$ for the *competences in answering questions* and finally $7.0 \pm 0,95$ for the *ability in reviewing students' ideas and concepts* and, as well, in *advising on possible future developments*. Only for the *ability to attract and generate interest* the students' score was relatively low: $5.40 \pm 1,15$. This value mirrors the existence of about 20% of students that, by default, tend to have a negative opinion on SWA activities, considered not useful for their future. Of course this is an issue that deserve attention and require adequate actions to counteract the contrastive attitude of this minority, possibly by mean of more personalized activities.

4   Future Work and Perspectives

Although the IoP experience described in this paper, could be considered overall a positive one, it represents only a first step toward SWA processes capable to contribute significantly to the increase of the learning ecosystems' smartness [13]. Students, in fact, identified also several aspects that could be improved in the future:

- better integration with the curricular activities;
- greater dilution of the "theoretical pills" within practical activities;
- denser relationships with companies and territorial stakeholders;
- reduction of the dead time during face to face revisions;
- availability of a more detailed planning since the beginning.
- The above indications suggest future directions to work on:
- acceptance of SWA activities by all schools' teachers that, hopefully, should contribute with their competences to the success of the process; a role of paramount relevance is certainly played by the internal tutors that should be trained in advance to stimulate: (a) students to respect the agreed schedule, (b) colleagues to collaborate;
- better design of the process that, ideally, should: challenge continuously the students involving them in practical activities, without disregarding the theoretical relevance to the overall process; not overlap and be in competition with curricular activities;
- development of a more effective networking action that is of paramount relevance not only for territorial development and social innovation but also, and overall, to train future active and participative citizens.

It is quite evident that SWA processes and activities can largely benefits from the *use of collaborative working/learning environments*, to make interact all players. However on this issue one has also to face some criticalities: standard working/learning environments offer obsolete or, at least, less efficient communication tools so that students tend to communicate quickly for example by WhatsApp or Facebook; to work collaboratively one could also use the suite of the tools created by Google; however the mentioned environments are not interoperable among them and are not interoperable with the learning environments; these latter, on the other, usually offer analytics that are not provided by social environments. Basically one has to choose between supporting efficiently working and communication processes or support efficiently learning and monitoring. The design of efficient SWA activities, in principle, have also to adopt strategies that mitigate this dichotomous situation.

Another aspect on which one should work in the future is the *certification of competences*. High quality SWA experiences are expected to provide also a certification of the competences that have been developed during the process, e.g. during the simulation of innovation processes. Students seem to be very interested in a certification that could be inserted in their curriculum, that can be communicated by mean of social networks and that can increase their level of employability. Our proposal to the students has been to implement a certification system based on the release of open badges. The proposal has been largely appreciated, on average 7.47 over 10. The main reasons have been: gratification for efforts, visibility in an e-curriculum, also in the perspective of start-up creation, enabling of new opportunities.

Certification is also an issue for internal tutors and teachers that expect their efforts to be recognized also in terms of long-life learning and competence transfer. Because of this, in the next future, we intend to work also on the definition of a certification system, open badge based, for tutors.

Finally another aspect that should be developed further is the monitoring and evaluation of SWA experience to collect also the opinions of other players and stakeholders (i.e. teachers, parents and territorial stakeholders) to

- compare their feeling with that of the students;
- support networking and sharing of responsibilities.

We do believe, in fact, that more meaningful SWA experiences can be realized only thanks to the involvement of the largest as possible number of members of the local communities rather than by standard and centralized procedure. Policies and centralized actions, at national and regional levels, should only define a flexible operative framework and the related guidelines (also to protect students), re-equilibrate resources not to leave anyone behind and help in disseminating and transferring best practices.

Acknowledgements. we are indebted with the SKILLAB school network represented by Prof. Arturo Marcello Allega, Director of the ITIS Giovanni XXIII, that was promoting together with ASLERD the PM2 project financed by USR Lazio.

References

1. Council Recommendation of 22 April 2013 on establishing a Youth Guarantee 2013/C 120/01. http://eur-lex.europa.eu/legal-content/EN/ALL/?uri=CELEX:32013H0426%2801%29. Accessed Feb 2018
2. Europe 2020 a strategy for smart sustainable and inclusive growth. http://eur-lex.europa.eu/legal-content/EN/TXT/PDF/?uri=CELEX:52010DC2020&from=EN. Accessed Feb 2018
3. Chatzichristou, S., Ulicna, D., Murphy, I., Curth, A.: Dual education: a bridge over troubled water, EU DG for internal polices - culture and education (2014). http://www.europarl.europa.eu/RegData/etudes/STUD/2014/529072/IPOL_STU(2014)529072_EN.pdf. Accessed Feb 2018
4. Reform of Vocational Education and Training in Germany. https://www.bmbf.de/pub/The_2005_Vocational_Training_Act.pdf. Accessed Feb 2018
5. Report on Vocational Education and Training 2015. https://www.bmbf.de/pub/Berufsbildungsbericht_2015_eng.pdf. Accessed Feb 2018
6. Terminology of European education and training policy-A selection of 100 key terms. http://www.cedefop.europa.eu/en/Files/4064_EN.PDF. Accessed Feb 2018
7. European Training Foundation (ETF): Work-based learning: Benefits and obstacles a literature review for policy makers and social partners in ETF partner countries (2013). http://www.etf.europa.eu/webatt.nsf/0/576199725ED683BBC1257BE8005DCF99/$file/Work-based%20learning_Literature%20review.pdf. Accessed Feb 2018
8. (2014). https://labuonascuola.gov.it/documenti/La%20Buona%20Scuola.pdf. Accessed Feb 2018
9. Giovannella, C.: School Work Alternance: from the state of the art to the construction of hypothesis for the future. In: Lifelong, Lifewide Learning (LLL), vol. 12, no. 28, pp. 123–134 (2016)

10. Zucca, G., Romagnoli, C., Italiano, L.: Le best practices realizzate dalle imprese nei rapporti con scuole e università (2016). http://www.un-industria.it/Prj/Hom.asp?gsAppLanCur= IT&gsPagTyp=21&gsMnuNav=01M:100,01L:1,01C:1,02M:0,02L:0,02C:1,&fInfCod=37 072&fPagTypOri=30. Accessed Feb 2018
11. Giovannella, C.: Schools as driver of social innovation and territorial development: a systemic and design based approach. IJDLDC 6(4), 64–74 (2016)
12. Giovannella, C.: An organic process for the organic era of the interaction. In: Silva, P.A., Dix, A., Jorge, J. (eds.) HCI Educators 2007: Creativity 3: Experiencing to educate and design, Aveiro, pp. 129–133 (2007)
13. Giovannella, C.: Participatory bottom-up self-evaluation of schools' smartness: an Italian case study. IxD&A J. 31, 9–18 (2016)
14. http://www.innovazioneinclasse.it/
15. Polaine, A., Løvlie, L., Reason, B.: Service Design: From Insight to Inspiration. Rosenfeld Media, New York (2013)
16. See for example. http://ischool.startupitalia.eu/education/53146-20160401-design-thinking-scuola and http://www.academia.edu/16783516/
17. Brown, T.: Design thinking. Harvard Bus. Rev. 1–10 (2008)
18. Martin, R.: The Design of Business: Why Design thinking is the Next Competitive Advantage. Harvard Business School Press, Boston (2009)
19. Giovannella, C.: Incubator of projectuality: an innovation work-based approach to mitigate criticalities of the Italian massive alternance scheme for the school-based educational system. IJDLDC 8(3), 55–66 (2017)
20. Giovannella, C.: Is complexity tameable? Toward a design for the experience in a complex world. IxD&A J. 15, 18–30 (2012)
21. Simon, H.A.: The Sciences of the Artificial. The MIT Press, Cambridge (1969)
22. Rittel, H.W.J.: Wicked problems. Manage. Sci. 4, 141–142 (1967)
23. Bauchanan, R.: Wicked problems in design thinking. Des. Issues 8(2), 5–21 (1992)

Interactive Learning in Smart Learning Ecosystems

Irene Merdian[✉], Gabriela Tullius, Peter Hertkorn, and Oliver Burgert

Reutlingen University, Alteburgstraße 150, 72762 Reutlingen, Germany
{Irene.Merdian,Gabriela.Tullius,Peter.Hertkorn,
Oliver.Burgert}@reutlingen-university.de

Abstract. The increasing heterogeneity of students at German Universities of Applied Sciences and the growing importance of digitalization call for a rethinking of teaching and learning within higher education. In the next years, changing the learning ecosystem by developing and reflecting upon new teaching and learning techniques using methods of digitalization will be both – most relevant and very challenging. The following article introduces two different learning scenarios, which exemplify the implementation of new educational models that allow discontinuity of time and place, technology and process in teaching and learning. Within a Blended Learning approach, the first learning scenario aims at adapting and individualizing the knowledge transfer in the course Foundations of Computer Science by providing knowledge individually and situation-specifically. The second learning scenario proposes a web-based tool to facilitate digital learning environments and thus digital learning communities and the possibility of computer-supported learning. The overall aim of both learning scenarios is to enhance learning for diverse groups by providing a different smart learning ecosystem in stepping away from a teacher-based to a student-centered approach. Both learning scenarios exemplarily represent the educational vision of Reutlingen University – its development into an Interactive University.

Keywords: E-learning · Interactive University · Learning methods

1 Introduction

Two changes make the development of Interactive Learning in Smart Learning Ecosystems necessary. (1) The German Higher Education System is characterized by a heterogeneous composition of students with different university entrance qualifications. Due to their different educational biographies, students do not show a homogeneous level of knowledge; furthermore, their access to course content and their individual learning methods are very diverse [1]. The varying degrees of knowledge and the very unequal study speed have a significant influence on the students' learning behavior and learning motivation. (2) Increasing digitalization can be observed in all areas of life and, consequently, has found its way into higher education. Hence, our students require the necessary skills to find their way in a digitalized society, the so-called 21st century skills.

Particularly, they have to face self-regulated learning (with respect to lifelong learning), collaborative learning and digital learning.

Both developments, the increasing heterogeneity of students in universities and the increasing importance of using digital technologies, open up new possibilities for universities but present new challenges at the same time. With regard to teaching and learning, heterogeneity and digitalization require a development from traditional to new learning environments. This was considered a major necessity in Horizon Report 2017, a report which presents annual trends and technology developments for higher education institutions. Blended Learning and collaborative learning are just some trends for the next years [2].

In the following, we will introduce two learning scenarios which take place in the Faculty of Computer Science at Reutlingen University. The first learning scenario represents a didactical concept which combines digital offerings and classical lectures in the basic modules of Computer Science. The second learning scenario is about a collaborative, interactive and web-based learning system, called Accelerator which allows learning and teaching to be spatially separated. Both approaches were developed separately, but they have been combined in the last semester to create new learning experiences in a richer learning ecosystem. They will be interleaved even more in the forthcoming semesters.

2 Background and Related Work

2.1 Reutlingen University as Interactive University

Reutlingen University with currently approx. 5800 students is a medium-sized university of applied sciences in South Germany. It offers 45° programs in five faculties: Applied Chemistry, ESB Business School, Engineering, Textile & Design and Computer Science. The university is aware of its social responsibility and is committed to equal opportunity and diversity.

Due to the German academic system, there are various ways to get accepted for a study program. At traditional universities, pupils enter university with A-levels, whereas at a university for applied science, students might, for example, have qualified for entrance by a (dual) traineeship or work experience. Hence, depending on the study program, we deal with highly diverse groups of students. Another dimension of diversity is the students' international background. Approx. 25% of the students are international or have a migration background. Many of the students are the first in their families who have the possibility to attend university. While this is characteristic for a university like Reutlingen University, which gives many students the option to build a very successful career, these students are not used to an academic environment. During the first semesters, the dropout rate is appreciably high because it takes time and effort to adapt to an unfamiliar environment.

Reutlingen University is strongly interested in dealing with these heterogeneous educational biographies. Our educational vision is to develop the Interactive University where students are encouraged to organize their development both from a subject point of view as well as from a personal point of view. We believe this will be the key for their future career and for our society; we want to mediate possibilities and methods for lifelong learning which enable our students to analyze problems and to come up with appropriate solutions. To achieve these objectives, we take e-learning to the next level. E-learning has been known and tried for more than 25 years, various didactical concepts and technical concepts have been examined [3]. While learners still need to learn, the role of educators is changing – teaching is no longer just imparting knowledge but more and more, communicating, collaborating, coaching and orchestrating the learning experience. Although developments cannot be foreseen with certainty, observations presented in reports like Higher Education [2] or Educause [4] imply this change.

Considering the future strategy of Reutlingen University, we think it is a major chance and challenge to develop and reflect upon new teaching and learning techniques enhanced by digital technologies. The overall aim is to enhance learning for heterogeneous groups by providing smart learning ecosystems. As many other researchers in the field, we change from a teacher-based approach to a student-centered approach.

In the process of developing the Interactive University vision, a group of interested colleagues developed a definition of interactive learning and teaching. The questions, which they posed themselves and others are, for example, how can teaching promote interaction? how do learners experience interaction? how can interaction achieve an added value in the learning processes? and how does our learning-and-teaching ecosystem look like? The basic idea of the concept of interactive teaching and learning is the advancement of a stronger self-regulated learning. Discussions have shown that for such a teaching and learning process the following conditions are necessary:

- Interaction requires a culture of experimentation and, consequently, a culture in which students are allowed to make mistakes.
- Interaction should be encouraged by using different media.
- In times of digitalization, a university must ask itself if there is an advantage to classroom teaching and how interaction in in the classroom can be promoted.
- Problem-based learning should be increasingly promoted and implemented as a didactical concept.
- Interactive design needs to be integrated in all basic courses.

2.2   Related Work

The importance of self-regulated and collaborative learning reverts to long existing teaching and learning theories [5, 6]. Moreover, different studies have already verified the importance of self-regulated learning to the academic performance [6–9]. At the same time, collaborative learning influences self-regulated learning. Regarding to learning outcomes, the higher the learning outcomes, the greater is the need for collaborative learning, even using digital media [10]. Computer based learning environments increase the opportunity for self-regulated learning but also present challenges to teachers and learners. Since it cannot be expected that learners bring the competencies like self-regulated, collaborative and digital learning when entering the academic higher education [11], the task of a university should be to create smart teaching and learning environments, which allow students, in addition to cognitive skills, to also develop, so called 21st century skills.

3 Learning Systems at the Faculty of Computer Science

At the faculty of Computer Science at Reutlingen University, we provide three under-graduate study programs "Media and Communication Informatics", "Medical Technical Informatics" and "Business Informatics" plus three postgraduate programs for currently around 850 students in total. At the faculty of Computer Science, less than 50% of the students enroll with A-level. The following two learning scenarios exemplify the implementation of new smart educational models allowing discontinuity of time, technology, place and process in teaching and learning as well as the already described aspect of interactive learning.

3.1 Curriculum 4.0

The project Curriculum 4.0, supported by Stifterverband and Carl-Zeiss-Foundation, is a teaching concept carried out in the Foundations of Computer Science course within the study program "Medical Technical Informatics". The following figure represents the didactical concept (Fig. 1).

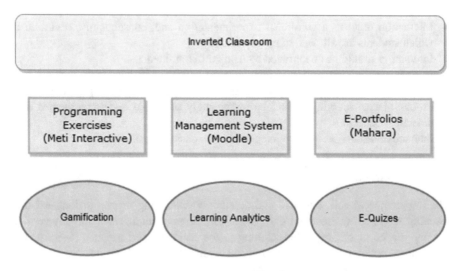

Fig. 1. Didactical concept of Curriculum 4.0

The didactical concept is based on the method of the Inverted Classroom. According to a Blended Learning approach, a combination of digital offerings and classical lectures, its focus is to adapt and individualize the knowledge transfer in Foundations of Computer Science by providing knowledge individually and situation-specifically. The concept integrates the teaching and learning tools used at the university: a learning management system, an e-portfolio and an in-house tool for online submission of programming tasks called "Meti Interactive". The exercises in the learning management system build on the idea of gamified approaches and pursue a motivational goal. Learning analytics allow the observation of the learning process and support students as well as teachers in the organization and planning of the learning process.

The benefits of Blended Learning are diverse. It allows a temporal flexibility of knowledge transfer, thus learners can autonomously decide on their learning time. Learning materials and diverse exercise tasks can be recorded and repeated at any time. The individualization of the depth of knowledge is another benefit. The learning content is grouped into different levels of subsequent difficulty. Each content block is only available to the students, if they already mastered the prerequisite topics. This allows the learner to acquire new knowledge linked to previous knowledge. Hence, an individual access to the content is possible. This satisfies the demands of different types of learners.

For the implementation of the Inverted Classroom, a didactical planning for the self-regulated learning phases but also for the presence phases is necessary. It requires a rethinking, away from the teacher-centered concept to a learner-centered environment. The learning environment is crucial for the fostering of self-regulated learning [11].

3.2   Accelerator

With Accelerator, a web-based learning system, we propose a technological base, a digital tool to facilitate learning communities, hence, a base for computer-supported collaborative learning (CSCL) and digital learning environments [4].

In an early article Etzioni et al. investigate into the communication of communities both face-to-face and computer-based [10]. Many of the points made there can be adapted to learning communities both onsite and offsite, e.g. break-out and reassemble, cooling-down effects and the like. It is well known that the gap between face-to-face and computer-based communication clues, like gesture and mimic, has to be overcome.

The Accelerator presents a digital tool which encourages and facilitates interactive learning and teaching. The idea behind is a virtual classroom metaphor, hence, collaboration at the same time (synchronic) at different locations [12]. Accelerator can be seen as part of a blended learning setting. Web conference systems are not new, numerous systems are on the market and used by companies to work location-independently (e.g. Adobe Connect, WebEx, TeamViewer and the like). These systems have been established and have been used in industry for collaboration, however, they mostly just fulfill the purpose of communication, therefore, replacing telephone calls by web conferencing. There seems to be little knowledge in how to use them in a meaningful manner. Collaboration in a computer based communication, i.e. virtual environment is different from collaboration in face-to-face situations. Means to enhance social presence and awareness have to be provided by the systems and have to be used in a meaningful way by the users [13]. Examples are the use of avatars and a number of informal awareness clues, like showing information about the state of the participants.

With Accelerator we have developed a system following the Xerox "eat your own dog food" idea, becoming users of the own system. Students develop their own virtual classroom which is used for lectures and seminars. From a technical perspective, Accelerator is based on web technology, thus no proprietary software is necessary and the virtual classroom can be accessed simply by using a web browser. The system provides clues on social presence and awareness and has a number of features one would expect from such a system, such as screen sharing, audio and video communication, chat, sharing and presenting files such as office documents, images or generally speaking pdfs as well as a collaborative editor and a collaborative whiteboard.

With Accelerator, we take the idea of interactive learning and teaching from CSCL to a next level: computer-supported-interactive-learning. Whereas in CSCL collaboration is met as a design challenge, our approach focuses on the interactive learning experience. We face the challenge to structure social interaction and mutual added-value for all members of the learning environment. From own observations and comparisons, we see a benefit of the interactive part in motivating users to take part in sessions. They can do so by annotating material online, by taking part in votes, by sharing and writing down thoughts, by finding consent etc. However, this means that the sessions have to be prepared well in advance, just like a good onsite seminar: which interactions will be used when, e.g. watching a YouTube video via Accelerator, stopping at a certain time and discussing the topics shown? What can be done if there is no reply to a question – how to deal with the lack of social clues? For interaction, we came up with a number of methods. As an example, if the session topic was to discuss components of the next generation digital learning environment (NGDLE) model [4] as described in Educause participants could be asked to augment the model given using virtual cards with their own ideas (shown in the figure below by the card with the Accelerator logo on) and to rate them. As a result, all participants can be active and have the chance to reflect on the given topic. Thus, the learner is in the center of activity (Fig. 2).

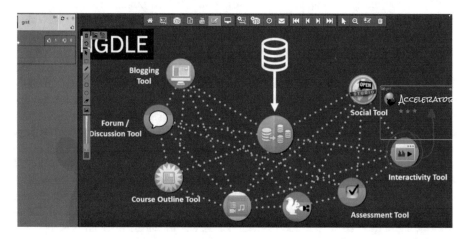

Fig. 2. Excerpt screenshot of Accelerator with augmentation of a given model, in this case the NGDLE

4   Conclusion and Future Work

Do Curriculum 4.0 and Accelerator solve all challenges of diverse student groups? Of course not. The Accelerator is a technical tool that has to be filled with well-thought-after content and didactical concepts. With the Curriculum 4.0 project the base is provided.

In first experiments, we used Accelerator during a presence phase of the Curriculum 4.0 course to allow remote students to participate in a teaching session. This was well received by the students, but it demonstrated that in larger groups the participation of students must be regulated to avoid distractions. We assume that smaller working groups working jointly on assignments between presence phases, which are supported by a tutor or lecturer, are more beneficial.

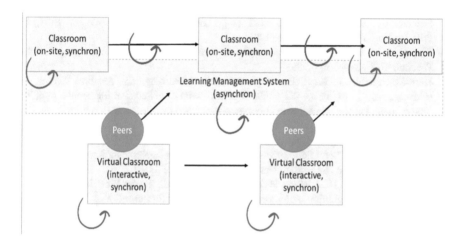

Fig. 3. DeLE – Digital (enhanced) learning environment with Curriculum 4.0 and Accelerator

The further development of the project Curriculum 4.0 is the design of self-regulated learning phases using the web-based learning system Accelerator (see Fig. 3) with constant feedback (see arrows). Self-regulated learning will be supported e.g. by special topic sessions in which students can gather in rooms and interact freely, working on projects at their own pace, while the teacher is available online for help. We will implement at least two of such sessions in the summer term 2018 for coding topics. The use of Accelerator could be above all to reach outcomes in higher learning goals, which require an increased exchange and discourse among students, within learning groups.

Acknowledgments. "Curriculum 4.0" is funded by Stifterverband and Carl-Zeiss-Foundation.

References

1. Dräger, J., Ziegele, F.: Hochschulbildung wird zum Normalfall. Ein gesellschaftlicher Wandel und seine Folgen. CHE, Centrum für Hochschulentwicklung gGmbH. Gütersloh (2014). Accessed 14 Feb 2018
2. Adams Becker, S., Cummins, M., Davis, A., Freeman, A., Hall Giesinger, C., Ananthanarayanan, V.: NMC Horizon Report: 2017 Higher Education. The New Media Consortium, Austin (2017)
3. Kahiigi Kigozi, E., Ekenberg, L., Hansson, H., Tusubira, F.F., Danielson, M.: Exploring the e-learning state of art. Electr. J. e-Learn. 6(2), 77–88 (2008)
4. Pomerantz, J., Brown, M., Brooks, D.C.: Foundations for a next generation digital learning environment: faculty, students, and the LMS. Research report. ECAR, Louisville, CO (2018). https://library.educause.edu/~/media/files/library/2018/1/ers1801.pdf. Accessed 14 Feb 2018
5. Bandura, A.: Social Foundations of Thought and Action: A Social Cognitive Theory. Prentice-Hall, Englewood's Cliffs (1986)
6. Zimmermann, B.: A social cognitive view of self-regulated academic learning. J. Educ. Psychol. 81(3), 329–339 (1989)
7. Himpels-Gutermann, K.: Selbstlernphasen und E-Learning. In: Armbrost-Weihs, K., Böckelmann, C., Halbeis, W.: Selbstbestimmt lernen – Selbstarrangements gestalten. Innovationen für Studiengänge und Lehrveranstaltungen mit kostbarer Präsenzzeit, pp. 103–116. Waxmann, Münster (2017)
8. Benz, B.F.: Improving the Quality of E-Learning by Enhancing Self-Regulated Learning. A Synthesis of Research on Self-Regulated Learning and an Implementation of a Scaffolding Concept. Technische Universität, Darmstadt [Dissertation] (2010)
9. Donker, A., de Boer, H., Kostons, D., Dignath-van Ewijk, C., van der Werf, M.: Effectiveness of self-regulated learning strategies on academic performance: a meta-analysis. In: Educational Research Review, pp. 1–26 (2014)
10. Würffel, N.: Gestaltung von Selbstlernphasen in Blended-Learning Kursen. Was gilt es zu bedenken? In: Armbrost-Weihs, K., Böckelmann, C., Halbeis, W.: Selbstbestimmt lernen – Selbstarrangements gestalten. Innovationen für Studiengänge und Lehrveranstaltungen mit kostbarer Präsenzzeit, pp. 125–134. Waxmann, Münster (2017)
11. Gerholz, K.-H.: Der Weg zu selbstregulierten Lernen als didaktische Herausforderung. In: Armbrost-Weihs, K., Böckelmann, C., Halbeis, W.: Selbstbestimmt lernen – Selbstarrangements gestalten. Innovationen für Studiengänge und Lehrveranstaltungen mit kostbarer Präsenzzeit, pp. 27–38. Waxmann, Münster (2017)
12. Etzioni, A., Etzioni, O.: Face-to-face and computer-mediated communities, comparative analysis. Inf. Soc. 15(4), 241–248 (1999)
13. Johansen, R.: Groupware: Computer Support for Business Teams. The Free Press, New York (1988)
14. Allmendinger, K., Kempf, F., Hamann, K.: Collaborative learning in virtual classroom scenarios. In: Cress, U., Dimitrova, V., Specht, M. (eds.) Learning in the Synergy of Multiple Disciplines. EC-TEL 2009. Lecture Notes in Computer Science, vol. 5794. Springer, Heidelberg (2009)

When Smartness of a Participatory Learning Ecosystem Should Not Be Interpreted as Mediation by Technology: Case-Study of Golbaf Town, Iran

Ali Maleki, Najmoddin Yazdi[✉] [ID], Milad Jalalvand, and Seyed Reza Tabibzade

The Research Institute for Science, Technology, and Industry Policy (RISTIP),
Sharif University of Technology, Tehran, Iran
{A.Maleki,Najmoddin.Yazdi}@sharif.edu, Jalalvand.Milad@ut.ac.ir,
Tabibzade_SR@gsme.sharif.ir

Abstract. Sustainable development is coined with ongoing social learning processes. As part of a sustainable regional development project in Golbaf town, Iran, development of a participatory community-based learning ecosystem (social learning) was soon found to be a requisite. This in turn was seen to be hampered by lack of social capitals, namely trust, self-confidence and participatoriness.

The interim results of the project indicate the followings as a way towards smartness (survivability) of a participatory learning ecosystem in developing contexts where mistrust, inactiveness and lack of confidence prevail: (1) Facilitating rather than doing by conveners, (2) learning by doing by citizens, and (3) gradual trust formation. It also questions suitability and survivability of highly technology-mediated learning ecosystems in such cases characterised by mistrust and lack of confidence.

The results suggest a progressive approach towards mediation of technologies. In fact, above socio-cultural barriers required us to proceed face-to-face for the regeneration of social capital in order to make the newly-born learning ecosystem survivable and embeddable by time.

Keywords: Sustainable development
Participatory community-based learning ecosystem · Mistrust
Lack of self-confidence

1 Introduction

Sustainable development is coined with plurality in understanding it. Thus, it is characterised with an ongoing social learning process [1]. Although, literature streams such as learning ecosystems, community-based learning (CBL), participatory learning and problem-driven learning should conceptually be linked with sustainable development, they are not yet in considerable contact with each other. They are still mostly entrenched in the classic stream of (formal) education and pedagogy.

© Springer International Publishing AG, part of Springer Nature 2019
H. Knoche et al. (Eds.): SLERD 2018, SIST 95, pp. 23–32, 2019.
https://doi.org/10.1007/978-3-319-92022-1_3

Seeing the inefficiencies of the government in creating sustainable employment, The Research Institute for Science, Technology, and Industry Policy (RISTIP) has started a first-of-its-kind sustainable regional development in Golbaf town, Iran, in 2016 with a focus on small home-based businesses. Having a little time passed, it was seen that a learning ecosystem is a prerequisite to sustainable employment. But, setting up a participatory informal learning ecosystem soon faced serious challenges. These challenges were found to be linked with social capital formation and active citizenship, which in turn were hindered by lack of self-confidence and trust.

The present paper tries to address this challenge by exploring linkages between social capital formation (with a focus on self-confidence and trust) and survivability of learning ecosystems (with a focus on participatory community-based learning) in a sustainable regional development context. Although, there are various studies on relationships between social learning and social capital, the role of social capital formation has not been studied much on sustainable development, urban regeneration or learning ecosystems (to point some: [2–4]).

In line with what is stressed on in sustainable development studies [5], the learning ecosystem studied here is not education-oriented as usual, but action-oriented (learning by doing), participatory and informal. It could be named "puzzling and powering" or "joint fact finding" [1]. In other words, it is a participatory community-based learning ecosystem. The developing context also seems important to the topic of social capital formation. These helps pinpointing the case and its generalisability.

Another contribution of the paper is that a technology-mediated learning ecosystem could not survive in such developing contexts where we still do not have trust, self-confidence, engagement and culture of participation among citizens. In contrast with the theme of the proceedings focused on technology-mediated learning ecosystem, it reconfirms what is stated in the preface of the 2nd SLERD proceedings 2017 that smartness does not simply mean technology-enhanced learning ecosystems but, rather, "learning ecosystems that promote the multidimensional well-being of all players of learning process and that contribute to the increase of the social capital of a region".

2   Background

We do not know what sustainable development exactly looks like. Thus, sustainable development is essentially conceived as an ongoing social learning process [6]. Such a knowledge on what is sustainable is heavily action-oriented (learning by doing) and contextual [1]. The action-orientedness means finding balances between what is desirable and what is feasible. This is referred to as "puzzling and powering" or "joint fact finding", too. Contextuality is itself defined spatially (place-based), temporally and societally [1]. The societal contextuality points to the participation and interaction of citizens in action-oriented learning process of sustainable development. In fact, while individual learning, learning in groups and organisations, and learning through innovation systems all take place in sustainable development, it is the learning via groups and interaction that is of foremost importance [1].

The above characteristics associated with learning towards sustainable development goals could best be matched with the term "participatory community-based learning". While community-based learning (CBL) is a well-known literature with a root in established literature of pedagogy and education, the participatory adjective departs it from education towards informal learning, which is emphasised in sustainable development context. Participatoriness implies an inevitable linkage with people, citizens and stakeholders. Thus, other literature streams could also be conceived as related, including civic engagement, civic formation, citizen participation (engagement), citizen learning and (pro)active citizenship. But, one could hardly find researches related to sustainable development within these streams. Therefore, these streams could not serve our purpose in studying participatory, informal, community-based learning with a sustainable development goal in mind.

Apart from community-based learning theory, there have also been other related theories nominated to build the study upon them, but which finally turned out not to be useful. One was socio-technical systems theory. Since technology was not of a focus in the studied ecosystem, the authors did not find it practicable. Another seemingly related theory was sustainability transitions. This theory was mostly found to be applied to socio-technical problems where technology's role is bolded. It usually has a concrete short-term or mid-term objectifiable state. But as it was aforementioned, sustainable development is coined with an ongoing social learning with an unknown end or even mid-term states, which makes it essentially a long-term phenomenon. Thus, sustainability transitions were found not suitable to the case's features. Social-ecological system theory was another candidate. Its focus on environmental ecology (linked with sociology concepts) was again found to be irrelevant to the studied case.

Furthermore, the topic is inherently interdisciplinary since it stems from a real-world societal initiative. It interrelates community-based learning, learning eco-system and informal participatory learning on one side, and sustainable development on the other side. Additionally, social capital formation has been proposed as a linked phenomenon. And lastly, technology mediation has been put forward as a failed agenda in the studied case. Such interdisciplinarity made authors unable to find similar previous works.

3   Methodology

This study is a qualitative and interventionist one. In interventionist research, researchers are directly involved in the real-time flow of events instead of observing at a distance or working with ex post facts [7]. The researchers were in fact doing the project during the last two years (2015–2017). They were in close contact with other actors, the community, actors and other researchers. In line with the nature of learning in sustainable development initiatives [1], the researchers did not intend to merely study an occurred phenomenon. Instead, research was continuously in agenda to give feedback to actions and designs. In other words, the project was at the same time design-oriented, action-oriented and explorative.

While participatory community-based learning approach was adopted to communicate the results of the project, no specific schematic model was assumed. This is because community-based learning is not the focus of the paper, but its linkage with social capital formation and technology mediation makes the contribution. No specific model was found for studying such linkage.

To measure trust and self-confidence as two elements of social capital formation, no survey or interview was conducted. The reasons are twofold. First, in an intervention research where researchers are directly involved in the real-time flow of events instead of observing them, they are of in-depth knowledge. In fact, instead of allocating some time to interview sessions, researchers spend much more time in a continuous manner with debated discussions and brainstorming during several months or years. Second, measuring trust and self-confidence was considered be counterproductive. Obviously, the efforts to gradually rebuild the trust and self-confidence among citizens during the last two years would be devastated by running a survey, or conducting interviews, asking them about their lack of self-confidence or lack of trust.

4   Golbaf Case

4.1   Case Introduction

Golbaf town has witnessed starting a sustainable regional economic development initiative in 2016 with the aid of The Resalat Network of Social Entrepreneurship, comprised of Bank of Resalat, Samin Entrepreneurship Development Group and an NGO active in social entrepreneurship. Initiating an informal learning ecosystem soon and seriously became attached to the economic agenda due to the socio-cultural challenges observed in the way towards sustainable development. It was hoped that the learning ecosystem could promote hope, self-confidence, participation, empowerment and teamwork in the community. Seeing the potential among children, they become the starting point and focus, while maintaining other groups in the agenda. The burden of setting up a learning ecosystem was put on The Research Institute for Science, Technology, and Industry Policy (RISTIP), which could be best characterised as a proactive multidisciplinary S&T research institute with a flexible structure. Table 1 provides a description of RISTIP researchers and local facilitators active in the project.

It should be noted that by local facilitators, it is meant those locals who have been heavily engaged with the project since its start. Of course, there are now lots of part-time volunteer local students, youths, teachers and trainers.

To set up a learning ecosystem, first seven local primary-school teachers were selected. It was planned to have interviews for the selection based on criteria of proactiveness, learnability, enthusiasm, social capital, social concern and influence on the local society. But, the Golbaf's Public Education Administrative claimed a more active role in this regard which resulted in a collaborative selection of teachers by the Administrative, supposedly based on above criteria. By now, it is turned out that the criteria were not met by all the selected teachers. Anyway, the most important criterion was proactiveness which should had been verifiable by previous engagement with societal problems. Its importance went back to the project focus of facilitation. This implied a need on delegation of roles from RISTIP researchers to local teachers as soon as possible. Table 2 provides a description of the selected teachers.

Table 1. Description of RISTIP researchers and local facilitators active in the Golbaf project.

Age	Gender	Education	Experience
RISTIP researchers			
24	M	Master of Social Policy	–
26	F	Master of Psychology	–
26	M	MBA	3 years
26	M	Master of Civil Engineering	2 years
27	M	Master of Entrepreneurship	2 years
32	M	PhD student of Science & Technology Policy	10 years
33	F	Bachelor of Spanish Language	–
34	F	Bachelor of Education Technology	–
40	M	PhD of Science & Technology Policy	10 years
40	M	PhD of Science & Technology Policy	10 years
Local facilitators			
30	M	Teacher and member of town council	–
30	F	Entrepreneur	–
38	M	Teacher and member of town council	–

Table 2. Description of selected teachers to participate in the training programme.

Age	Gender	Occupation	Teaching experience
24	F	Primary school teacher	2 years
37	F	Primary school teacher	10 years
39	M	Primary school teacher	8 years
41	F	School manager	15 years
44	F	School manager	20 years
45	F	Primary school teacher	26 years
49	F	Primary school teacher	22 years

The seven selected teachers participated in a compact training programme at two of the best schools (Mizan and Ruyesh Schools) of the capital city, which are known for adopting modern participatory models of education and training. Back to home, they were enthusiastic and felt ready to start implementation of their ideas. These teachers started close groupwork with RISTIP researchers and three proactive local facilitators.

Table 3 provides the list of activities and events of the project so far. The first initiative was a social movement called "Movement of My Golbaf" which has been supported by a dedicated website, Telegram channels, mottos and banners throughout the region. The first event was "Sentence, Idea and Image", in which citizens were asked to send images, sentences and ideas about their town to get voted. This event received more than 300 images, 59 sentences and 47 ideas for the Movement of My Golbaf. Bank of Resalat funded the prizes to the three best ideas, sentences and images, plus 5 winners by chance.

Table 3. Activities and events of Golbaf project.

Activity title	Type	GenderTime span
Sentence, Idea and Image	Competition	Sep 2016–Dec 2016
Collaborative recognition of the region (reports)	MovementInitiative	Sep 2016–Feb 2018
Hundred Games- Hundred Joys	Competition; Event	Oct 2017–Dec 2017
Sustainable development and social entrepreneurship	Workshop	Sep 2016
Movement of My Golbaf	Movement	Sep 2016–…
Sustainable development and social innovation	Workshop	Nov 2016
Tree planting ceremony (Planting 116 trees)	Event	Mar 2017M
Resilient Economy Festival	Event	Apr 2017F
Nature cleanup program	Event	Jul 2017F
Thousand Games Festival	Competition; Event	25–26 Aug 2017
Development of Golbaf's poetry and fiction collections for children and teenagers with the participation of locals	Initiative; Movement	Sep 2017– …F
Urban beautification movement	Movement	Feb 2018–…

The next event was a festival to know herbs of the region for healthcare purposes. Periphery to the herbal festival, a game festival named Hezar-Bazi (Thousand Games Festival) was co-designed by the teachers and researchers to provide a collective gaming experience for children alongside the participation of their parents. This festival was run in four rounds, each absorbing about 200 participants. With the aid of 4 RISTIP facilitators (researchers) and 40 local volunteers, 60 games were co-designed for children and their parents for a two-day festival. The festival was done in 5 stations, including balancing, focus and accuracy, joy and excitement, art and imagination, and consultation stations. The first three stations contained competitive games, while art and imagination station covered painting, face painting, handwork and pottery. The consultation station distinctly provided a space for parents to ask an invited consultant their doubts and questions about children. The teachers, local volunteers and organisers were devoted to not buy toys as much as possible, but to build them, co-design the space and games, and arrange the whole festival.

Nature cleanup program was an environmental event for children to clean the town for the first time. It was recently extended to an urban beautification movement. There have also been some other continuing efforts regarding learning of mothers, children and youth, participatory gaming, teamwork, home-based business and setting up social events. Every event and initiative were advertised via news websites, social network channels, posters and the official website of "Movement of My Golbaf".

4.2   Facilitating Rather than Doing

The participatory community-based learning ecosystem of Golbaf could be characterised by three pillars. The pillars are facilitation, learning by doing, and participation of citizens. The first pillar indicates that researchers and facilitators should limit their roles to facilitation as much as possible, with the highest degree of delegation. This would necessitate a proactive role for citizens. Although limiting the role of government in learning ecosystems to facilitation rather than implementation has been stressed much and is not a new thing, e.x. [8, 9], here the newness is about limiting NGOs and research groups. In developing countries with a resource-based economy, citizens are historically used to ask central governments as well as NGOs just for funds without being engaged in development processes. In fact, expectations of such inactive citizens are confined to formal and financial contributions totally on the burden of outsiders, e.x. setting up physical educational spaces with some borrowed teachers from outside of the region.

In line with a mere facilitation role, RISTIP's members decided to pale the presence of Resalat Bank as a backing funder in order to shift the attitude of the citizens toward a temporary volunteer NGO without much finance. This was hopefully thought to lead to capability building by the hands of themselves rather than looking for a source of short-term loans and funding. Furthermore, every task and design have soon been passed onto local citizens to stimulate their learning and proactiveness in a longtime process. For example, in setting up Thousand Games Festival, all the burden of the design and administration was passed to locals after a brief training on principles and best practices. But of course, coaching and answering questions remained a central task of RISTIP researchers throughout the process.

4.3 Learning by Doing

The second pillar of Golbaf's learning ecosystem is a focus on learning by doing. Learning by doing was totally odd to the citizens such that many did not believe in it as a learning, innovation or even a value-added effort. They expected a formal physical class or educational trip to major cities with some predefined exams. It was then aimed to help them discuss and deal with their emotions, economic and social problems of the region and their family problems. Therefore, it could be said that this pillar has not been yet accepted much in the minds of citizens.

The learning ecosystem could also be described as an effort towards community-based social innovation. This could be an alternative to market-led territorial development [10]. Recalling that innovation is essentially a learning and that learning is much more inclusive than development of new products and services, social learning would inevitably be intertwined with social innovation. Social innovation could be characterised by the two pillars of learning by doing and design thinking, while learning by doing is also a trending approach in modern learning ecosystems [10, 11].

4.4 Citizens' Participation

The third pillar of the learning ecosystem was citizen participation and participatory community-based learning. Citizen participation has been bolded as a paradigm shift in city services and more generally in learning ecosystems in ICT-based [12] and other regional development projects [13, 14]. But having proactive citizens could be named as the most challenging aspect of the Golbaf project. In fact, the culture does not support volunteer involvement of citizens, especially when facing an ecosystem set up by outsiders. It should be noted that so far, we were only able to get women and children engaged, but still no considerable achievement for involvement of men.

5   Survivability of Learning Ecosystem: Roles of Social Capital and Technology Mediation

Sustainability development is intertwined with ongoing social learning process [1]. Such a learning ecosystem could be characterised as participatory, community-based, action-oriented (learning by doing) and informal. Thus, proactive citizenship becomes a core element in achieving participatoriness and collectiveness. Citizen participation is, inter alia, closely linked with social capital of a community. While these linear arguments seem rational and obvious, it was not the case for us until several months passed since the start of the overall project when we found social capital formation as an essential element for the survivability of learning ecosystems in Golbaf town. The social capital elements missing from the community were trust and self-confidence.

5.1   Social Capital: Mistrust

Trust constitutes a vital factor in social capital formation in learning ecosystems, urban regeneration and sustainable regional development [16, 17]. On the other hand, mistrust among citizens has been highlighted in regional development studies [18]. The resistance of Golbaf's citizens to participate in the learning ecosystem was found to be related to their mistrust to the organisers and to the success of such new informal programme. This was started by doubting the intentions and funding source of the RISTIP team, suspecting if they are pursuing a business agenda. Even, there was some mistrust and sense of unfairness to the selection of the two proactive locals and the teachers. The mistrust was interwoven with a sense of injustice in selecting winners of the events, selection of the teachers to be trained and prioritisation of problems. It was also evident in their pessimistic perception of feasibility of the learning ecosystem and the projected outcomes.

Such a mistrust was gradually tackled by our oversensitivity on fairness and transparency of the societal and design processes and also by adopting similar living standards, such as traveling by train instead of airplane, using local means of transportation or sharing foods. It should also be noted that trust building was much successful among women and children compared with men and male youths. In addition, the parallel community-based economic development programmes really helped and reinforced the process of social capital formation.

5.2   Social Capital: Lack of Self-confidence

Lack of self-confidence has also been confirmed as a key reason for unwillingness of citizens to commit to continuous learning and participation, especially when values of a community do not support inclusion and lifetime learning [19]. It was evident at the start of the project that a lack of self-confidence prevents citizens to get involved in discussions, brainstorming and criticizing each other's ideas. Talking about regional problems and criticizing presented ideas and solutions was completely odd to the citizens. As an example, the selected teachers were asked to have an oral presentation at the end of the training programme in the capital city of Iran. None of them were willing to or able to present. It is worth mentioning that the hypothesised relationship between technological literacy and unwillingness to participation were intuitively rejected since all, including those technologically or academically literate, depicted the same behavior. As another fact, in brainstorming sessions, nobody ever raised a question or did a criticism over our suggestions and nobody used to accept a responsibility for shared works. When asking them why, they confirmed that they fear to not be able to do tasks. Thus, the participatory learning ecosystem also targeted building up self-confidence. By now, one can see sparkles of hope in regeneration of self-confidence among the citizens, e.x. by actively participating in discussions, brainstorming, criticizing others' ideas and implementation of their ideas.

5.3 Technology Mediation

The case was so far characterised by mistrust and lack of self-confidence of citizens. These pointed to the necessity of social capital formation as a prerequisite to a viable community-based learning ecosystem. They also resulted unwillingness of citizens to participate in development and learning processes. Inability and unwillingness to participate posed a serious question on timeliness and survivability of a technology-mediated learning ecosystem. In fact, it was soon found that we need face-to-face human-centred communications to stimulate trust building and capability building (mainly regeneration of self-confidence). It should be accompanied by maintaining a focus on learning by doing. This is not what could be achieved by technology-mediated learning ecosystem without sufficient human-centred interactions and learning. As an instance, we could point to the serious difficulties faced with current simple technology-mediated systems during the last two years, namely participation of citizens in social network channels and supplying content to websites. Therefore, while simple technology mediums are to some extent in use, the researchers highly agree that turning the learning ecosystem, even a part of it, into a technology-mediated or a distant-learning one would not work now. A gradual adoption of technology-mediated learning systems is thus advised up to when there is enough social capital supporting them.

The case also demonstrated that a learning ecosystem by doing could be conceived as an ecosystem with boundaries highly blurred with diverse economic, social, gaming, household, religious and cultural activities of a comprehensive sustainable regional development programme.

References

1. Loeber, A., van Mierlo, B., Grin, J., Leeuwis, C.: The practical value of theory: conceptualising learning in the pursuit of a sustainable development. In: Social Learning Towards a Sustainable World, pp. 83–98. Wageningen Academic Publishers, Wageningen (2007)
2. Callois, J.-M., Aubert, F.: Towards indicators of social capital for regional development issues: the case of French rural areas. Reg. Stud. **41**(6), 809–821 (2007)
3. Iyer, S., Kitson, M., Toh, B.: Social capital, economic growth and regional development. Reg. Stud. **39**(8), 1015–1040 (2005)
4. Harriss, J., De Renzio, P.: POLICY ARENA: 'Missing link' or analytically missing? The concept of social capital. Edited by John Harriss. An introductory bibliographic essay. J. Int. Dev. **9**(7), 919–937 (1997)
5. Wals, A.E.: Social Learning Towards a Sustainable World: Principles, Perspectives, and Praxis. Wageningen Academic Publishers, Wageningen (2007)
6. Tilbury, D., Cooke, K.: A national review of environmental education and its contribution to sustainability in Australia: frameworks for sustainability. Department for the Environment and Heritage, and Australian Research Institute in Education for Sustainability (2005)
7. Jönsson, S., Lukka, K.: Doing interventionist research in management accounting. University of Gothenburg, Gothenburg Research Institute GRI (2005)
8. Goyal, S., Sergi, B.S.: Creating a formal market ecosystem for base of the pyramid markets-strategic choices for social embeddedness. Int. J. Bus. Glob. **15**(1), 63–80 (2015)
9. Pace, R., Dipace, A., di Matteo, A.: On-site and online learning paths for an educational farm. Pedagogical perspectives for knowledge and social development. Rem-Res. Educ. Media **6**(1), 39–56 (2014)
10. Moulaert, F., Nussbaumer, J.: The social region: beyond the territorial dynamics of the learning economy. Eur. Urban Reg. Stud. **12**(1), 45–64 (2005)
11. Rizzo, F., Deserti, A., de Pous, M.: Social Innovation Community EU Project. Deliverable 4.1. Report on SIC learning principles and processes (2017)
12. Mealha, Ó.: Citizen-driven dashboards in smart ecosystems: a framework. Interact. Des. Archit. **31**, 32–42 (2016)
13. Zago, R., Block, T., Dessein, J., Brunori, G., Messely, L.: Citizen participation in neo-endogenous rural development: the case of LEADER programme. In: 6th EAAE Ph.D. Workshop, Co-organized by AIEAA (Italian Association of Agricultural and Applied Economics) and the Department of Economics of Roma Tre University (2015)
14. Markkula, M., Kune, H.: Making smart regions smarter: smart specialization and the role of universities in regional innovation ecosystems. Technol. Innov. Manag. Rev. **5**(10) (2015)
15. Dondi, C., Aceto, S., Proli, D.: Learnovation Foresight Report. Foresight Report HAL Id: hal-00592999 (2009)
16. Chang, W., Cha, M.: Government driven partnership for lifelong learning in Korea: a case study of four cities. Int. J. Lifelong Educ. **27**(5), 579–597 (2008)
17. Hibbitt, K., Jones, P., Meegan, R.: Tackling social exclusion: the role of social capital in urban regeneration on Merseyside—from mistrust to trust? Eur. Plan. Stud. **9**(2), 141–161 (2001)
18. Schicklinski, J.: Civil society actors as drivers of socio-ecological transition?: Green spaces in European cities as laboratories of social innovation. Working Paper No. 102-THEME SSH. 2011.1.2-1 (2015)
19. Peirce, H.J.: The dynamics of learning partnerships: case studies from Queensland. Queensland University of Technology (2006)

Smart Schools with K9 Student Opinions: The Aveiro José Estêvão Case

Óscar Mealha[1(⊠)] and Fernando Delgado Santos[2]

[1] University of Aveiro/DigiMedia Research Centre, 3810-193 Aveiro, Portugal
oem@ua.pt
[2] Agrupamento de Escolas José Estêvão de Aveiro, Aveiro, Portugal
delgado@aeje.pt

Abstract. Smartness has been advocated with a strong bias, almost exclusive, on technology's characteristics and extension of implementation in diverse ecosystems. The research process reported in this paper takes place in the school ecosystem and considers all its stakeholders in a bottom-up approach to nurture the knowledge needed to understand the relation that is established between educational community and school. This approach does not exclude other top-down models and approaches to understand the school ecosystem, in fact, it supports the complementary need to have a correlation of both perspectives to develop on the community - school relation. The reported work contextualizes and explains the smart learning ecosystem strategy and describes the research method used to engage and inquire stakeholders concerning their wishes, interests and needs in the school ecosystem. The smart school questionnaires used to inquire the educational community's stakeholders will be described considering structure and nature of closed and open-ended questions that are used. The research procedure, data processing and analysis will be shared in the context of the pilot study that took place last spring 2017 in the José Estêvão Aveiro school cluster, with 7–9th grade students, n = 81. The potential for co-design of technology-mediated solutions (APPs/services) will also be discussed in the context of the qualitative opinion of these particular stakeholders.

Keywords: School smartness · K9 · Co-design · Communication
Educational community

1 Introduction

The "smart" concept is being used in extremely diverse ways and contexts, usually associated to the innovative characteristics of information technology or telecommunication's characteristics and functionalities in specific environments. This research highlights that this smartness concept can be highly potentiated when directly coupled and sustained on people-centred design [1]. A process supported by service design thinking methods [2] to promote social innovation and transformation. The SLERD 2018 conference theme "The interplay of data, technology, place and people", that hosts this publication, considers the core elements of this research: (i) the nuclear opinion of "people" - school stakeholders; (ii) school as institution and specific "place" and context; (iii) quantitative and qualitative data to understand diagnosis and subjective opinion from

© Springer International Publishing AG, part of Springer Nature 2019
H. Knoche et al. (Eds.): SLERD 2018, SIST 95, pp. 33–44, 2019.
https://doi.org/10.1007/978-3-319-92022-1_4

the educational community actors, the school's stakeholders and last but not least, (iv) one of the purposes of the application of the smart school questionnaires, to identify possibilities of co-designing "technology" mediated solutions to reduce the gap between educational community and school. The relation of an educational community and school is normally considered a priority in any place and in whatever stakeholder perspective you look. Each of the community's actors usually consider that an optimal relation benefits the excellence of the educational process of all students.

Historically, school has been studied by special interest groups [3], specialists supported by specific data gathered at schools (student's performance) and from some school stakeholders, typically top-down studies conducted by specific government policies and strategies. The research reported in this paper complements on these traditional approaches and supports its methodological approach on a bottom-up strategy, the educational community's stakeholders. Other studies, have been conducted on a same methodological basis in university campuses [4, 5] and recently extended and adapted for the school ecosystem in Rome's school clusters [6]. The concept of smartness is nurtured with the direct engagement of each of the study's participants, within a People-Centred Design process [1], initially with closed and open-ended question surveys and later in the research process, during the co-design phases, with interviews and focus groups sessions. This paper reports on the first stage concerning questionnaire supported inquiry, qualitative data analysis and potential for service design in the school cluster. The questionnaire structure and questions where adapted and localized in the Portuguese school context from previous pioneer work developed in Rome, Italy, by Giovannella [6]. The questionnaires have two main theoretical references, Maslow [7] Csíkszentmihályi [8], and although other recent studies [9, 10] can be found using human engagement strategies shaped by people's incentives that Barroca et al. call (2013) "*Personal Drivers - wishes, needs and interests, WINs*", these approaches resiliently contextualize into Maslow's [7] "Theory of Human Motivation" and at some point also into Mihaly's [8] Optimal Experience, Flow theory. The questionnaires are shaped by Maslow's dimensions: (i) Basic Needs; (ii) Security, (iii) Socialization; (iv) Social Capital, (v) Self-Fulfilment and (vi) Mihaly's Flow/Satisfaction.

2 Method

Research focuses on one of the six possible types of stakeholders, the K9 students of José Estêvão School Cluster at Aveiro (AEJE), Portugal, with n = 81 participants, 38,3% from 7th grade, 29,6% 8th grade and 30,9% form 9th grade. Gender is represented with 49,4% female and 50,6% male. Data from this pilot study was gathered during May 2017.

The study was authorized by the Ministry of Education of the Portuguese Republic and by the Director of AEJE school cluster (No. 0576100001) which included the validation of the study's procedures, instruments and ethical issues. Authorization from parents for under-18 year participants and a consent form for all participants was of utmost importance and required rigorous planning to comply civic classes' timetable that were used to apply the questionnaires within the school's context and time.

The 5–9th grade class student questionnaires based on research developed with ASLERD members [5, 6, 11] is structured in 13 parts directly related with 6 (B-G) Maslow/Mihaly dimensions:

A. (i) Socio-demographic data (3)
B. Basic Needs/Resources | (ii) Infrastructure (4), (iii) Environment (2), (iv) Food (1);
C. Security | (v) Security (3);
D. Social Capital | (vi) People and Space (11);
E. Socialization | (vii) Sociability (6), (viii) Interaction with Family (2), (ix) Social and Territorial Interaction (3);
F. Self-Fulfilment | (x) Educational Process (4), (xi) Personal Development (2), (xii) Communication (2) and
G. Flow | (xiii) Satisfaction (2),

32 closed, 10 open questions, 14 are related pairs of closed and open-ended questions. These B - G dimensions are directly coupled to Abraham Maslow's Theory of Human Motivation - THM [7] and Csíkszentmihályi's Theory of Optimal Experience - Flow [8]. Theories that also sponsor the scientific and strategic people-centred design methods used in this research process to idealize, design and develop technology mediated products and services to optimize the relation of educational community and school as institution and place [12]. Table 1 systematizes the relation of both reference theories, questionnaire's structure/dimensions and corresponding closed and open-ended questions. Some of the closed questions also included a comment (open text box) to understand the reason of valorisation given by each participant.

Table 1. Relation of reference Theories, Smart Questionnaire structure and Questions.

Theory/Dimension	Open-ended (OQ) and related Closed Questions (CQ)
1. THM, Basic Needs/Infrastructure, Equipment	CQ1. On a scale of 1 to 10 (very low 1–10 very much), how much do you like the space in your school - classrooms, labs, auditoriums, sports facilities, WC, public areas? (w/comments) OQ1. What other spaces would you like to have and for what? CQ2. On a scale of 1 to 10, how much do you like the equipment of your school - classroom furniture, computers, sports equipment, etc.? OQ2. What additional equipment would you like to have at your disposal and for what?
2. THM, Basic Needs/Environment	CQ3. On a scale of 1 to 10, indicate to what extent you are careful at school with the environment (garden care, recycling, etc.)? OQ3. In your opinion, what environmental problems do you find in school?

(continued)

Table 1. (*continued*)

Theory/Dimension	Open-ended (OQ) and related Closed Questions (CQ)
3. THM, Basic Needs/Food Services	CQ4. On a scale of 1 to 10 what is your rate of the canteen service? (w/comments)
4. THM/Security	CQ5. With a scale of 1 to 10, to what extent do you feel safe in your school? CQ6. With a scale of 1 to 10, to what extent do you feel safe out of school? OQ4. In your opinion, what are the main security problems within your school?
5. THM, Socialization/People & Space	CQ7. With a scale of 1 to 10, indicate how well you feel at your school? (w/comments) CQ8. With a scale of 1 to 10, indicate to what extent you get along with your colleagues? (w/comments) CQ9. With a scale of 1 to 10, indicate how well is your relationship with the teachers? (w/comments) CQ10. With a scale of 1 to 10, indicate to what extent you discuss with your colleagues? (with comments) CQ11. With a scale of 1 to 10, how much do your classmates value you for your results in school, sports or other activity? (w/comments)
6. THM, Socialization/Interaction w/Family	CQ12. With a scale of 1 to 10, indicate to what extent would you like your parents to collaborate with the school? OQ5. Please give some examples of activities you would like your family to do at school?
7. THM, Social and Territorial Interaction	CQ13. With a scale of 1 to 10, indicate to what extent you consider that the school develops a relationship or visits the city? OQ6. What initiative would you like the school to organize? CQ14. With a scale of 1 to 10, indicate to what extent you would like to have a space to communicate through the computer with your colleagues and teachers from home? (w/comments)
8. THM, Self-Fulfilment	CQ15. With a scale of 1 to 10, indicate how satisfied you are with the challenges and opportunities offered by your teachers (exchanges with other schools, contests, participation in competitions, etc.)? OQ7. What other things do you propose or do differently?
9. THM, Self-Fulfilment/Educational Process	CQ16. With a scale of 1 to 10, how much do you like the daily organization of your school? (w/comments) CQ17. On a scale of 1 to 10, to what extent do you think the school's actions have been useful and well developed to help improve the performance of students with learning difficulties? (w/comments)
10. Flow, Satisfaction	OQ8. What do you like about this school? OQ9. What don't you like about this school?

3   Findings

Qualitative data analysis, originated by open-ended questions and comments directly coupled to quantitative, closed questions, (10 point Likert scale, Minimum 1 - Maximum 10), has two purposes in this research, one is related with the validation of the smart school questionnaire and the other to gather data that might represent a potential for design. This instance of inquiry can explicit the expectations and needs of the study's participants and thus constitute an opportunity to design technology mediated solutions, APPs, WebAPPs, etc. Figure 1 represents a holistic view of the average score (μ) and standard deviation (SD, σ) given by the n = 81 participants in each closed question. The radar axis number identifies the closed question and the legend establishes the question - study dimension relation.

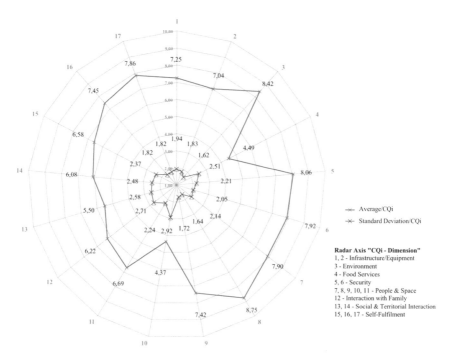

Fig. 1. Average (μ) and SD (σ) of each closed question related with open-ended questions/comments.

The open-ended questions' data analysis, as systematized in Table 2, did not reveal any suggestion to increase, or reduce, the smart inquiry dimensions or critique to any specific question. This can be considered a preliminary indicator of "validation evidence" for the smart school questionnaire, as is, from these specific stakeholders.

Table 2. Analysis of term occurrence in the context of each question.

Question	Occurrence of terms in open answers and comments
CQ1. On a scale of 1 to 10 (very low 1-10 very much), how much do you like the space in your school - classrooms, labs, auditoriums, sports facilities, WC, public areas? (54 comments)	Positive: social space (1); place to play (1); classroom/lab., equip., AC, +organization (17); love school (15)
	Negative: canteen, +tables/chairs (4); canteen food (2); outdoor sports facilities, rain (1); classroom/lab., cold, equip. maintenance, projector (8); door/locks, WC paper, small locker room (11); outdoor painting (1); rain shelter (5); wifi (1); room 16 small (1); small socializing room, missing chairs, more socializing space (5); closed corridors (1); outdoor pavement (1); short class break (2); int/ext. garden care (2)
OQ1. What other spaces would you like to have and for what? (67 answers)	Positive: school space is OK (8)
	Negative: videogames (6); study room (1), different space 4 basket & football (3); socializing room, socialize w/older students (21); +sports facilities, cultural, skate-park, table tennis (13); wifi (4); vegetable garden (1), rain shelter (4); PC, AV projectors (1), green space, garden bench (6); canteen (1); movie room (1); WC w/doors (4); +Labs (4)
OQ2. What additional equipment would you like to have at your disposal and for what? (68 answers)	Positive: nothing missing (8)
	Negative: PC, videogames, entertainment (8); AV projectors, computers (7); +bar space (3); +canteen space, +Tables (4), use uniform (1), +sports equip., balls, nets, bikes, badminton racquets, skate, roller-skates, mini-golf (14p/14); +net/wifi (17); tablet/student (7); +equip special needs (1), music in school (3); school radio (2); +sports facilities (1); heating (2); socializing space (2); +microwave (1); table soccer (1); +TV (1)
OQ3. In your opinion, what environmental problems do you find in school? (68 answers)	Positive: none (15)
	Negative: +dustbins in/outside (4); careless people (9); litter floor (27); garden care (8); recycling bins, more (4); mud w/rain (2); social outdoor areas (1)
CQ4. On a scale of 1 to 10 what is your rate of the canteen service? (66 comments)	Positive: cooks, staff (4); don't eat at canteen (10); bar food (1); food, service OK (4); price (1)
	Negative: food waste (1); food quality, process, flavour (50); more tables, chairs (2); queue (3); social place (1)
OQ4. In your opinion, what are the main security problems within your school? (68 answers)	Positive: none (19)
	Negative: staff vigilance, corridor, gates (37); Bullying, bad treatment (2); scared older students (1); video-cameras needed (1); smoking (2)
CQ7. With a scale of 1 to 10, indicate how well you feel at your school? (63 comments)	Positive: like the school, feel well (15); staff (12); students (23); teachers (12); security (4)
	Negative: don't like the school, cold, garden, broken Tables (6); teachers (2); staff (4); older students, students (5); security (3); place to keep backpack (2)

(continued)

Table 2. (*continued*)

Question	Occurrence of terms in open answers and comments
CQ8. With a scale of 1 to 10, indicate to what extent you get along with your colleagues? (63 comments)	<u>Positive</u>: people who like me (7); all colleagues (35); educated, sense of humour (4); tolerant (8)
	<u>Negative</u>: don't like my classmates (3); tease me (5); many groups (4)
CQ9. With a scale of 1 to 10, indicate how well is your relationship with the teachers? (61 comments)	<u>Positive</u>: Supportive, comprehensive, patience, care (35)
	<u>Negative</u>: not friendly (8); fast teaching (5); unfair, punishment (12); too much rigor (9); lack of rigor (1)
CQ10. With a scale of 1 to 10, indicate to what extent you discuss with your colleagues? (58 comments)	<u>Positive</u>: mild discussions, music (13); important issues, school, politics, ideas, civilized (18); I'm friendly, no conflicts (16)
	<u>Negative</u>: I don't like them (2); think they're important (1)
CQ11. With a scale of 1 to 10, how much do your classmates value you for your results in school, sports or other activity? (59 comments)	<u>Positive</u>: support, critic (33); like me (1); class OK (4)
	<u>Negative</u>: I don't reveal my results (4); they don't care (4); 2nd thoughts, critic (6)
OQ5. Please give some examples of activities you would like your family to do at school? (60 answers)	<u>Positive</u>: visit the school (1); parent association (5); games w/children, theatre, sports, contests (10); assessment meeting w/parents (1); assist to classes (3); lunch w/teachers and students (4)
	<u>Negative</u>: none, need to work (27)
OQ6. What initiative would you like the school to organize? (64 answers)	<u>Positive</u>: study visits, national/international, London, Italy (38); solidarity walk (8); activities in the park (3); theatre (1); community project w/city (2); bike ride (2); visit sports professionals (2); visit firefighters Dep. (1)
	<u>Negative</u>: none (2)
CQ14. With a scale of 1 to 10, indicate to what extent you would like to have a space to communicate through the computer with your colleagues and teachers from home? (61 comments)	<u>Positive</u>: communicate colleagues, group work (9); exam preparation (7); homework support (26); videogames (2);
	<u>Negative</u>: not necessary (15); students should socialize at school (2); already have email, social media (7)
OQ7. What other things do you propose or do differently? (59 answers)	<u>Positive</u>: study (1); outdoor classes (3); classes (1); mobility K9 students (7); +sports time (2); tests, homework (2); study visits (12); videos that teach (1); contests (4); group work (3); relevant talks (1)
	<u>Negative</u>: nothing more (8); less homework (1)

(*continued*)

Table 2. (*continued*)

Question	Occurrence of terms in open answers and comments
CQ16. With a scale of 1 to 10, how much do you like the daily organization of your school? (59 comments)	Positive: Organization OK, morning better (20); teachers are organized (2); break org. OK (1); classes timetable OK (5); more time for lunch (6)
	Negative: staff needs org. (1); need a socializing student room (2); too many books everyday (5); more org. corridors (3); improve classes timetable (10)
CQ17. On a scale of 1 to 10, to what extent do you think the school's actions have been useful and well developed to help improve the performance of students with learning difficulties? (52 comments)	Positive: teachers help OK (1); lab is used (1); project "nests" (10); study support classes (4); tutoring OK (2)
	Negative: no opinion (3); project "nests" n-ok (2)
OQ8. What do you like about this school? (75 answers)	friends (18); public space, informatics room, sports facilities, arts (12); school environment (16); school organization (23); gardens (1); teachers (11); class breaks (2); bar food (2); big school (4); study visits (1)
OQ9. What don't you like about this school? (73 answers)	classes (2); teachers unfairness (6); school (1); more tables for canteen (4); food (21); time for lunch (2); school board (2); more study visits3); staff (10); students (1); 2 students/computer (2); security & surveillance (2); litter floor (2); no opinion (7)

The analysis of the open-ended questions and comments directly coupled to the closed questions mention in Table 2 and corresponding descriptive statistics in Fig. 1, is structured according to the smart inquiry dimensions, shown in Table 1, as follows.

3.1   Infrastructure

This section of the questionnaire was scored relatively high with an average score of $\mu = 7,25$ ($\sigma = 1,94$) for CQ1 and $\mu = 7,04$ ($\sigma = 1,83$) for CQ2. The most mentioned issues in this dimension were related with more and diverse space in school at the canteen, for socializing (indoors and outdoors), more sports facilities and also require improvement in management, namely of sports equipment. The improvement of outdoor and indoor gardens was also mentioned for socialization activities.

3.2   Environment

The closed question related with school environment, CQ3, got a mean score of $\mu = 8,42$ ($\sigma = 1,62$) with a strong opinion on excellent school environment. As an improvement contribution, the large majority of participants mention litter on the floor (27/68, 39,7%), some reason on that and mention the lack of dustbins as the main cause, but a few comments also express that cleaning staff are not responsible for this unpleasant situation, but instead, the students that do not have the correct behaviour and litter the indoor corridors and outdoor socializing space and gardens, as mentioned in the following excerpt *"We sometimes find litter on the floor. Not that the cleaning staff don't do their work but because students aren't careful"*.

3.3 Food

The food service closed question (CQ4) has the second lowest scored mean $\mu = 4,49$ but with one of the highest standard deviation $\sigma = 2,51$. The open-ended comments directly related to the closed question reveal that a clear majority of participants (50/66, 75,8%) consider the canteen's food is very bad and many explain why. Badly processed, cooked food, sometimes raw food due to lack of cooking time, this excerpt of the data is an excellent example of this type of comment, *"I don't think the canteen should serve gourmet food but they should pay more attention to the cooking process. I've been surprised by raw food"*. Although the question only concerned food, data also reveals that a big number of participants had the need to highlight that the canteen staff are extremely nice and 15,2% (10/66) mention they don't eat at the canteen.

3.4 Security

The closed questions related with security issues got the third highest mean scores of the smart questionnaire, CQ5 $\mu = 8,06$ ($\sigma = 2,21$) and CQ6 $\mu = 7,92$ ($\sigma = 2,05$), so its clear a great majority of participants feel secure at school and around the school facilities. Although participants feel very secure they suggest improvements, 54,4% (37/68) still express that the school entrances should have some kind of surveillance. Whenever this is mentioned some comments reveal students are frightened that anyone can go into the school facilities without any kind of security control or screening. The second biggest cluster of opinions, 30% (19/68), just reveals that the school has no security problems, they feel secure! This bipolar majority of opinions is curious and future research has to be done on this topic to better understand the cause of such distinct opinions in one same institution and place.

3.5 People and Space

This dimension has 5 open-ended questions with open-ended comments, the richest qualitative cluster of this questionnaire. Most participants mention they like the school because people are friendly, students, staff and teachers and they feel secure in a good environment. A huge majority, 85,7%, with the highest mean score on CQ8 $\mu = 8,75$ ($\sigma = 1,64$), mention they have an excellent relation with colleagues and socialization is extremely good. A few mention 1 or 2 situations of students that don't like them but besides that all's fine. Concerning the relation with teachers, 80,3% mention the teachers are excellent professionals scored on CQ9 $\mu = 7,42$ ($\sigma = 1,72$), very rigorous and sometimes not very tolerant and fair (punishment) dealing with some situations. Most of the participants reveal that they don't regularly have discussions with colleagues, also expressed in closed question CQ10 $\mu = 4,37$ ($\sigma = 2,92$), occasionally they do discuss some things but it depends of the topic, sometimes important issues, others just "childish" matters. A majority of the students, 64,4%, mention that their work or sports activities are valued by their peers regularly and sometimes reinforce the good relation they have in class. Question CQ11 confirms these comments with a mean score of $\mu = 6,69$ ($\sigma = 2,24$).

3.6   Interaction with Family

When questioned on how the family could interact with the school, opinions are divided with a minor advantage for the cluster of participants that just say they don't want (45%, 27/60) their parents at school at all, CQ12 $\mu = 6,22$ ($\sigma = 2,71$). The other second biggest cluster of opinions would like to have parents directly involved with school activities (40%, 24/60), sports, having meals at school and even participating in some classes.

3.7   Social and Territorial Interaction

The great majority would like to have the school organize more study visits (64,1%, 41/64), namely abroad, some even mentioning the city. Participants clearly low rate this school activity in CQ13 $\mu = 5,50$ ($\sigma = 2,58$), opinions depict students want more visits. The second major cluster of opinions requires the organization of outdoors social support activities, with public visibility. Students revealed that a technology mediated communication service could be interesting to discuss with teachers and colleagues difficult topics but besides that a major amount of participants just mention that they don't need it. Some even mention that when they do need such an instrument to virtually communicate with someone, concerning school, they use social networks. The mean score of this issue in CQ14 $\mu = 6,08$ ($\sigma = 2,48$) isn't, curiously, very high considering student age and predictability for new media usage, most probably because this technology is announced to mediate human relations for the educational process' purpose. Some opinions get it clear that *"students and teachers should socialize at school"*.

3.8   Self-fulfilment | Personal Development and Educational Process

When inquired on the personal development topic several clusters of suggestions occurred related with outdoor activities at school, interchange programs with other schools and activities with older students. Participation in more study visits would be good. The question on self-fulfilment CQ15 has a moderate average score $\mu = 6,58$ ($\sigma = 2,37$) so more can be done.

A majority of students (47,5%, 28/59) consider the school to be very well organized but even so, mention some things should be improved. They mention that their school backpacks are heavy and could have some way of not carrying around such a lot of books. Others mention that their school timetable could be improved to have more free time and extra activities during the class breaks. The support to students with learning difficulties, are also mentioned to be efficient because the students with special needs start getting better grades. Teachers are referred as key players in the, specific need, educational process and support to students that reveal needing some kind of special support, CQ16 $\mu = 7,86$ ($\sigma = 1,82$).

3.9 Satisfaction

What participants most like in school is its organization (21,3%, 16/75) and environment (30,7%, 23/75), then opinions are divided between human relationship and indoor and outdoor space. The people they like to meet at school most are their colleagues but occasionally some comments go to the teachers and school staff. The sports facilities and equipment are the most mentioned alongside some remarks to indoor spaces. What these 7–9th grade stakeholders dislike at school is clearly the canteen and its food (34,2%) alongside the canteen space because it's too small to cope with efficient feeding time for all students during lunch. Outdoor equipment to shelter from rain is mentioned by a big number of participants and in third place some students mention human relations with some teachers and staff as not being pleasant, sometimes. In some of these situations it's understood that this comment is related to lack of tolerance and rigorous requests of staff and teachers.

4 Potentiating Co-design

The qualitative data analysis, checked with corresponding closed question scores, reveals some recurrent issues that have to do with quality assessment of service (canteen service) and organization (canteen lunch hours). Although some participants, in some occasions, mention human relations issues, the potential for technology mediated solutions considering these stakeholders opinion is related with civic behaviour concerning litter on the floor, space organization and food service, namely canteen space organization, queue efficiency and food quality/process. The next phase of this research is to share these findings with the K9 stakeholders and discuss and validate this possibility of co-designing a webAPP with a gamification approach to deal with these issues, civic behaviour, organization/management of school indoor/outdoor space and food service. This product should also include basic canteen information and management functionalities like weekly food menu and acquisition of meals and socialization room management. On a daily basis the access to the canteen could also be integrated in the smartphone APP without need to have a specific card for this purpose. These and other technology mediated characteristics and design strategies, to fulfil the solutions for some of the problems identified in the smart school questionnaire data still need discussion and validation with these stakeholders.

5 Conclusions

The school stakeholders considered in this paper, the K9 students (7th–9th grade students at AEJE school cluster) did not refute any question or smart school inquiry dimension, nor did they suggest any other issue not predicted in the smart school inquiry structure. This suggests that these particular school stakeholders accepted this smart school questionnaire as reasonable to gather a diagnosis (closed question scores) and subjective opinion (open-ended questions and comments) concerning school as an institution and place. The smart school questionnaire qualitative data revealed potential opportunities for co-designing solutions to mediate a better relation between the K9 stakeholder and school. The most prominent subjective opinions are related with civic behaviour, space organization and management (socialization) and the canteen food service and lunchtime management, with a potential to follow on with a co-designed contribution for this issue. The food supply service is of an external school partner so a technological mediated support service could be an add-on to increase efficiency of this service. Future work will relate these K9 stakeholders' opinions with the other five stakeholders' opinions to find a common co-design framework for this school cluster.

Acknowledgements. Research reported in this paper was also supported by the grant SFRH/BSAB/128152/2016 (Fundo Social Europeu and Portuguese financial resources from the Ministry of Science, Technology and Higher Education - MCTES). A special acknowledgment to all the teachers, parents and students at the school cluster Agrupamento de Escola José Estêvão de Aveiro, Portugal, 7th–9th grade 2016/2017, that shared their opinion and disposed of part of their time for this research.

References

1. Norman, D.A.: The Design of Everyday Things. MIT Press (2002)
2. Stickdorn, M., Schneider, J.: This is Service Design Thinking. BIS Publishers, Amsterdam (2011)
3. Pereira, H., Mil-Homens, P., Rocha Pinto, M.L., Lourtie, P.: As Universidades Públicas estão em época de exames. Eleição da Melhor Universidade Pública. Diário de Notícias, Lisboa, Portugal (2001)
4. Galego, D., Giovannella, C., Mealha, O.: An investigation of actors' differences in the perception of learning ecosystems' smartness: the case of university of aveiro. Interact. Des. Archit. **31**(1), 19–31 (2016)
5. Giovannella, C., Andone, D., Dascalu, M., Popescu, E., Rehm, M., Mealha, O.: Evaluating the resilience of the bottom-up method used to detect and benchmark the smartness of university campuses. In: Smart Cities Conference (ISC2), 2016 IEEE International - Improving the Citizens' Quality of Life, pp. 1–5 (2016)
6. Giovannella, C.: Participatory bottom-up self-evaluation of schools' smartness: an Italian case study. Interact. Des. Archit. **31**(1), 9–18 (2016)
7. Maslow, A.: Theory of human motivation. Psychol. Rev. **50**(4), 370–396 (1943)
8. Csíkszentmihályi, M.: The Psychology of Optimal Experience. Harper Collins (1990)
9. Oliveira, Á., Campolargo, M.: From smart cities to human smart cities. In: Proceedings of the Annual Hawaii International Conference on System Sciences, pp. 2336–2344, (2015)
10. Barroca, J., Brito, D., Campolargo, M., Concilio, G., Ferreira, V., Martires, P., Rizzo, F.: MyNeighbourhood Concept (2013). http://my-neighbourhood.eu/. Accessed 15 Jan 2018
11. Galego, D., Giovannella, C., Mealha, O.: Determination of the smartness of a university campus: the case study of Aveiro. Procedia Soc. Behav. Sci. **223**, 147–152 (2016)
12. Mealha, O.: Citizen-driven dashboards in smart ecosystems: a framework. Inter. Des. Archit. J. IxD&A **31**, 32–42 (2016)

Smart Learning and Territory

Unravelling the Role of ICT in Regional Innovation Networks: A Case Study of the Music Festival 'Bons Sons'

Paula Alexandra Silva[✉], Oksana Tymoshchuk, Denis Renó, Ana Margarida Almeida, Luís Pedro, and Fernando Ramos

University of Aveiro/DigiMedia, Campus Universitário de Santiago, 3810-193 Aveiro, Portugal
`{pags,oksana,denis.reno,marga,lpedro,fernando.ramos}@ua.pt`

Abstract. Since the beginning of the century two thirds of the Portuguese territory is threatened by desertification and the decline of economic activities. To face such trends, regions need to implement innovative strategies that leverage on the endogenous resources of the territory to foster economic recovery and to promote entrepreneurship, creativity, smart learning, and innovation. This paper reports on the study of the Bons Sons music festival as an example of an initiative developed in a low-density population area that mobilized their endogenous territorial resources to promote growth and economic development. The case study, which was based on the descriptive and qualitative analysis of a semi structured interview with the artistic director of the festival, aims at understanding the role of digital technologies in the process of regional innovation. The article contributes with an analytical view of community networks mediation practices and offers a set of tips and recommendations for the effective creation and consolidation of mediation strategies, community networks, and learning ecosystems that foster regional innovation.

Keywords: Regional innovation · Mediation · Community networks
Digital technology · ICT · Digital innovation

1 Introduction

The development of digital technologies, such as social software and social networks, led to new forms of collective action, online social interaction (synchronous or asynchronous), and sharing of collaborating spaces, as well as the production, distribution, and aggregation of information in online environments [1]. This Information and Communication Technologies (ICT) context allows for the dissemination of information through connections between individuals with common interests, promoting the collaborative construction of knowledge [2] and the development of smart learning ecosystems.

The flexibility of ICT tools coupled with its ease of use and wide availability have produced changes in the way social interactions are established [3]. Because of their proliferation, ICTs have been used in numerous forms of human activity, where regional innovation is not an exception. Low-density regions are particularly interesting

© Springer International Publishing AG, part of Springer Nature 2019
H. Knoche et al. (Eds.): SLERD 2018, SIST 95, pp. 47–61, 2019.
https://doi.org/10.1007/978-3-319-92022-1_5

ecosystems to analyse within this scope. According to Castro, Santinha, & Marques [4], the contribution of digital technologies to regional innovation unfolds in three main axes. The first supports economic development, facilitating, for example, the interaction of agents with scientific and technological systems, the flow of information and interaction among agents, and the production of knowledge. The second axis sustains social and regional cohesion, allowing for greater equity in access to services by citizens, facilitating the interaction of social networks, promoting equal economic-social and ecological opportunities, and contributing to a greater connection among both rural and urban centres and their influencing areas. Finally, the third axis supports administration, allowing for greater efficiency in its internal operation tasks, facilitating the provision and exchange of information with citizens and agents, promoting participation in policy-making and decision-making processes, and improving the community's overall quality of life through a more intelligent access to and use of resources and people's knowledge.

Digital technologies play a role in the processes of social and regional innovation, yet little is known about its specific role in maintaining, expanding, and mediating community networks. This will be studied in this paper, as not only do we want to understand the role of ICT in regional innovation networks but also we want to explore the usefulness and validity of a research instrument designed to research the subject.

This study is part of a larger research effort taking place under the umbrella of CeNTER – Community-led Networks for Territorial Innovation – an interdisciplinary program based at the University of Aveiro, Portugal, that aims to identify the different tools, from policy to management, and digital media, which best enable the valorisation of territorial resources in the central region of Portugal. To investigate the role of ICT in regional innovation networks we will resort to a specific case study of regional innovation - the Bons Sons music festival – that emerged at the core of a previously lethargic rural village – the village of Cem Soldos in Tomar, Portugal. Specifically, we analyse an interview conducted with the artistic director of the festival. Bons Sons is the largest and most diverse festival dedicated to Portuguese music that since 2006 took it upon itself to bring life back to the village of Cem Soldos, while disseminating Portuguese music and artists, and providing participants with an immersive experience in the life of the village that reveals itself as a rich informal learning ecosystem. The Bons Sons initiative is one of the pilot studies that will lay the grounds to the investigation of a larger set of initiatives in the context of the CeNTER program.

2 Related Work

Unravelling the role of ICT in constructing a scenario of regional innovation and promoting smart learning ecosystems, requires developing a sound understanding of concepts such as innovation, mediation, community, and social networks. It is also important to get a grasp at how technological competences and propensity for technology adoption can be measured. These topics are covered in this section.

2.1 Innovation, ICT Mediation, and Community Networks

Innovation is a constant concern in contemporary society, where innovative formats refer to creative and effective solutions to existing problems. Schumpeter [5] considers two types of innovation, one referred to as technological innovation, which is of a material nature, and another that concerns the innovation process, which is not tangible. Digital technology innovations are situated somewhere in-between those two types of innovation, as they may not materialize into a physical product. In the contemporary media ecosystem it is frequent to find both types of innovation in a single experience. The Bons Sons Festival is such an example as it encompasses both types of innovation. Furthermore, the case of Bons Sons is aligned with the ideas of Schumpeter [5], for whom intangible values are decisive in social innovation processes, and despite being situated in material concepts, technology is built from new processes and logics. The same author also argues that valid radical innovations are those that trigger revolutionary changes and truly transform society and the economy [5, p. 74].

Directly related to innovation we find the construction of social networks. In the contemporary media ecosystem, some theorists [6–8] also refer to social networks as social means, as this is a process of building networks of users with common interests in mediatized environments. However, in this research, we consider the concept of networks as the most appropriate, as it accommodates not only processes built in mediatized environments, but also from physical approximations. This is in-line with the concepts of Castells [9, 10], who considers social networks as the construction of relationships through common interests, whether they take place in digital environments or not.

The concept of mediation defined for this study unfolds from the classical one proposed by Martín-Barbero [11], for whom mediating means setting an equidistant common point of reference between two parties, which allows both parties to establish some type of interrelationship. This means that mediations are communication strategies that allow human beings to represent themselves and their surroundings, in a process of producing and exchanging meaning. Scolari [12] has complemented the concept of mediation with the one of hypermediation, to address the extended scale and power of mediation when it takes place in digital environments. According to Jenkins, Ford, and Green [13], the relationship between social actors in mediatized environments exists since the Web 2.0, and mediation is crucial for an effective digital revolution to take place:

> A persistent discourse of "Do-It-Yourself" media (Lankshear and Knobel 2010), for example, has fuelled not only alternative modes of production but also explicit and implicit critiques of commercial practices. Meanwhile, the rhetoric of "digital revolution" and empowerment surrounding the launch of Web 2.0 has, if anything, heightened expectations about shifts in the control of cultural production and distribution that companies have found hard to accommodate. [13, p. 53].

Castells [9] argues that the evolution of digital technology and social media has significantly influenced the development and management of communities in processes of regional innovation. This requires an understanding of what constitutes a community. A community implies that a group of people shares an interest, concern, and motivation on a given subject, interacting actively to deepen their knowledge in that subject [14, 15]. Nowadays, communities are no longer limited to specific places, instead they have become geographically wide-ranging and the main reason for their formation are their common interests, rather than simple regional/local affinities [16].

The concepts discussed in this section frame the study reported in this paper. The understanding of these concepts is crucial not only in guiding the selection of the case study, but also in defining the scientific concerns underlying it. This mapping of ideas serves then as a conceptual research map, which can be adopted in future studies on similar subjects.

2.2 Understanding and Measuring the Use of ICT

Interaction and collaboration digital tools are essential for regional innovation communities, often composed of voluntary members who do "not work together on a day-to-day basis" [17, p.10]. Creating and fostering online communities is a challenge and requires the development of a sense of belonging and of strategies that support participants in fitting in and working together for a mutual purpose [18]. It is also important to note that, as Wenger et al. [17] state, "Good technology in itself will not a community make, but bad technology can sure make community life difficult enough to ruin it." (p. 9). Therefore, the process of providing real communities with technology needs to focus on the solution and be designed to respect its circumstances, aspirations, members, and activities.

In an attempt to better understand users' needs and to clarify the factors that contribute to, or inhibit, the use of digital technologies, Davis [19] developed the Technology Acceptance Model (TAM). TAM defines that perceived usefulness and perceived ease of use determine an individual intention to use a system, where a positive perception regarding ease of use positively influences the perception of usefulness. Building upon Davis's work, Parasuraman and Colby [20] developed the Technology Readiness Index (TRI), a scale that measures people's propensity to embrace and use cutting-edge technologies for accomplishing their goals. This scale segments users according to a gestalt of mental motivators and inhibitors that collectively determine a person's predisposition to use new technologies [21]. In 2015, Parasuraman and Colby [22] published TRI 2.0 that assesses both overall technology readiness and individual technology readiness components, such as optimism, innovativeness, discomfort, or insecurity. The TRI scale has been widely used and translated into a number of languages [21], and studies confirm that the scale contributes both to the theory of technology acceptance and to the understanding of how people react to new high-technology products [23–26].

As already mentioned, digital technologies are key drivers of innovation, but not everyone has the knowledge, skills, and attitudes to use digital technologies in a critical, collaborative, and creative way. The Digital Competence Framework for Citizens (DigComp) was developed with the objective of identifying digital knowledge and the digital needs citizens have in their social and personal life. DigComp 2.1 structures competences in five areas [27]: (i) Information: Browsing, searching, and filtering information; Evaluating Information; Storing and retrieving information; (ii) Communication:

Interacting through technologies; Sharing information and content; Engaging in online citizenship; Collaborating through digital channels; Netiquette; Managing digital identity; (iii) Content-creation: Developing content; Integrating and re-elaborating; Copyright and Licenses; Programming; (iv) Safety: Protecting devices; Protecting data and digital identity; Protecting health; Protecting the environment; and (v) Problem-solving: Solving technical problems; Expressing needs & identifying technological responses; Innovating, creating and solving using digital tools; Identifying digital competence gaps.

DigComp 2.1 further presents eight proficiency levels, which are methaphorically compared to the competencies of a swimmer. Each proficiency level is described in relation to the complexity of the tasks' problems and the level of autonomy, together with the description of the competence in terms of learning outcomes [28]. First published in 2013, DigComp has become a reference for many digital competence initiatives [27]. It has also been accepted as a framework for e-skills indicators in Eurostat's household survey in the 2015 [29] and used in a number of other studies (for example, [30–32]).

In this study, TAM, TRI, and DigComp 2.1 were used to shape the interview guide (further detailed in Sect. 4) used to conduct the study with Bons Sons. TAM and TRI were used as a reference to develop the interview questions to uncover the relationship of the community with technology, while DigComp 2.1 was used to identify the community's digital competences in the areas of 'Information' and 'Communication'.

3   The Case of the 'Bons Sons' Festival

Bons Sons is a music festival that takes place since 2006 in Cem Soldos, a village of 600 inhabitants, located in Tomar, in the centre of Portugal. A local association –Associação Cultural Sport Club Operário de Cem Soldos (SCOCS) – organizes the festival, which motto is "Come Live the Village". The festival offers a platform for the dissemination of Portuguese music, featuring a diverse lay-out of artists, from famous musicians to emergent projects. Furthermore, the festival offers participants an opportunity to have a wide community experience, including its social and cultural dimensions.

As we can read in the Bons Sons website[1], the organizers of the festival also aim to support the regional development of the village by retaining younger members of the community and by boosting local economy and everyone in the village is involved in doing so. When attending Bons Sons, visitors are invited to immerse themselves in the village, its people, spaces, and rituals. Local people open their houses and offer them as accommodation to visitors with whom they often share meals. Local places, such as the school and the church, are used to hold the concerts. Aware of the positive impact of the festival in developing the region, the whole village is involved in welcoming the festival, in an event that is not only a musical festival, but also a rich and engaging social, cultural, and touristic experience.

Since it began, Bons Sons has transformed a local folk celebration into a renowned cultural event, which has grown ever since its first edition, from receiving 20 thousand

---

[1]   http://www.bonssons.com/en/bons-sons/ .

visitors in 2006 to 35 thousand in recent years[2]. Bons Sons's merits go beyond the increase in the number of visitors, and has been nationally and internationally recognized by several awards, such as: 'Most Sustainable Festival' in 2014[3], 'Best Line-out' and 'Best Hospitality and Reception' in 2017[4], and 'Be Green' in 2017[5]. According to the Bons Sons website, the festival has featured over 200 bands and 237 concerts. Since 2016, information about the festival is also available through a mobile app 'Bons Sons'.

Bons Sons has become an exemplar festival not only for its remarkable cultural impact, local development, community engagement, but also for its environmental, economic, social, and regional sustainability. The extraordinary impact and history of Bons Sons makes it a compelling case study. To thoroughly understand how the festival is organized and the strategies behind its success, may give way to replicating those strategies in other remote areas with similar potential.

4   Research Design

A case study methodology was chosen because this is a useful methodology when the purpose of the research is to perform an intensive detailed examination and to gain a rich and in-depth understanding of a given subject [33]. Within the case study, we resorted to a semi-structured interview, because this method allows for the rich collection of data, while still lending the researcher the necessary flexibility to direct the dialogue with the interviewee in such a way that the researcher can effectively capture the interviewee's point of view, goals, and concerns [33].

The first step of the research was to create an interview guide (Portuguese version available on request). The interview was organized in three main topics, each contributing to the understanding of an important aspect of how a community gets organized and the role ICTs play in that process. The first topic of the interview was intended at uncovering community organizational aspects, such as structure, leadership, definition of strategic goals, as well as the flexibility and proactivity of the community network. The second aimed at revealing community mediation dynamics, such as those related to collaboration, cooperation, interaction, and relationships between members. Finally, the third looked into the technologies the community avails of to further identify the tools, features, and platforms the community uses, as well as the readiness and competences of use. Table 1 presents an overview of the interview guide by topic and the dimensions and metrics observed under each topic as well as the studies used to frame each of the concepts involved.

---

[2] https://tinyurl.com/y7dqzc29 (Accessed Feb 9 2017, only available in Portuguese).
[3] https://tinyurl.com/y7tjba74 (Accessed Feb 9 2017, only available in Portuguese).
[4] https://tinyurl.com/yb9x433p (Accessed Feb 9 2017, only available in Portuguese).
[5] http://www.bonssons.com/i/2017/PR-BS-20170627.html.

Table 1. Structure of interview guide by topic and its dimensions and metrics

Topic	Dimensions and Metrics
Community Organization	Definition of strategic goals – defined collectively, alone [34]
	Leadership – shared, formal, informal [18, 34]
	Organizational structure – teams and network creation [18, 34]
	Flexibility and proactivity of the community network – creativity and openness to new ideas and initiatives [35]
Mediation Dynamics	Interaction – community connectivity, synergies, and cohesion [18, 36]
	Relationships between members – trust and level of compromise [18, 34, 35]
	Collaboration – shared construction of knowledge [34, 36]
	Cooperation – shared problem solving [34, 36]
Digital Technologies	Technology use –characterization of the community regarding the use of technologies [17, 18, 37]
	Technology readiness– based on TRI [22]
	Digital competences– based on DigComp [28]
	Digital tools and resources – tools and resources used to support activities [17, 18, 37]
	Platforms – Platform-level considerations [17, 18, 37]

Once the interview guide was finalized, a first contact by email was made with the interviewee - the current festival director - inviting and inquiring about his availability, to which we got a positive reply. At the day of the interview, and before initiating the interview protocol, the researcher introduced herself, explained the goal of the study, and obtained the interviewee informed consent. A record of the interview was kept and later transcribed verbatim for further analysis. The researchers performed the qualitative analysis of the interview collectively and the process of analysis and interpretation was guided by the themes of the interview guide.

5   Data Analysis and Results

The case study of Bons Sons was based upon a semi-structured interview with the current managing director of SCOCS and artistic director of Bons Sons (since 2006), a 34-year-old Industrial Designer, from Cem Soldos. The interview lasted about 40 min, was recorded, and transcribed verbatim. This section reports on the analysis and results of the interview. The dimensions of analysis are aligned with the ones used to structure the interview guide: community organization (structure, leadership, definition of strategic objectives, flexibility and proactivity of the community network), mediation dynamics (collaboration, cooperation, interaction, and relationships among members) and technology (tools, features, and platforms).

5.1   Community Organization

The community of Bons Sons is organized in teams, each sharing specific responsibilities in the project for the duration of one year. Each team has a coordinator and there are six areas of coordination: service production, stage handling, computing, finance, communication, and volunteers. With the exception of the artistic director, executive director, and financial director, all other members of the project are volunteers. Leadership is informal and shared, as this allows for the involvement of other community members. As the results of the project go towards the community and not individuals, everyone naturally takes on their responsibilities, which reflects a horizontal organization and relationship among members.

The Bons Sons Festival has well-defined strategic goals, all of them aiming to improve the quality of life of the people in the village. One of the goals is to demonstrate that there is a place for villages in contemporary society, i.e. as stated by the director in the interview, "the city is not the future or the country the past", both the city and the country should live in harmony. In parallel, the project also seeks to build a sense of union in the local community and to develop the involvement of the community in the project.

Those who created and first envisioned the festival structured a system that encourages both proactive and flexible attitudes in the participants. "Each one always draws up their own objectives, according to the artistic direction's greater design, which is my responsibility, where programming is also my responsibility, but increasingly the team knows what Bons Sons stands for", the interviewee said. Despite being voluntary work, human resources are greatly valued and, in the words of the interviewee, "every person is a key piece to keep the machine working". Work satisfaction is another important aspect and "Work has to be enjoyable, it has to be good. Because if it's not good, there's no point", the interviewee declared.

5.2   Mediation Dynamics

Mediation dynamics such as collaboration, cooperation, and interaction between members stand out in the interview. In this study, we consider the concepts of collaboration and cooperation presented by Dillenbourg and shared by Antikainen:

> "In cooperation, partners split the work, solve sub-tasks individually and then assemble the partial results into the final output. In collaboration, partners do the work together". (Dillenbourg 1999, p. 8 apud [38], p. 45).

Setting out from these concepts, which are similar and often taken to be synonyms [38], we highlight the division of tasks in the Bons Sons Festival project as something collaborative, while external participation ensures cooperation. The interviewee asserts "we have a lot of partnerships, that we go looking for or they come to us, but it's the only way... we don't have money, we've got ideas, we've got some time, less and less, and we have human resources; it's our currency." Recognizing human resources are Bons Sons most valuable resource, the festival organizer is aware of Bons Sons' responsibility and impact in the training and life of their volunteers, and the interviewee remarks "... by participating in this type of projects, one is embedded in a logic of informal learning. The model of empowerment is very serious and complements academic training. We have kids in Cem Soldos who have defined themselves academically because of the work they are doing at Bons Sons, and we have other people who have defined Bons Sons, because of their knowledge and contribution to the festival."

However, the interviewee underlines that the project "needs an injection of capital to be able to carry out its plan to study how the example could be spread in the other villages, not as a festival, but the skeleton of the organization".

Collaboration, i.e., when everyone works together towards a single result, is evident throughout the whole interview, however this aspect particularly stands out at one instance. Every year the group makes a self-assessment and holds a reflection session with all participants at the end of the event to identify what aspects needs to be improved and if training is needed to address those. The interviewee says "training takes place every year… and this year again we'll strengthen other training areas such as the stage management and technical support, because we are renewing teams and we feel the new members need this training." Still on the importance of the reflection session, the artistic director of Bons Sons states that allows "people to understand the project, and to feel and see themselves in the project". That feeling of belonging is fundamental for renewed collaboration and to foster unity. "We say we're a village that believes, and by believing, we get things done", are the interviewee's final words.

Collaboration does promote interaction among the members and the feeling of belonging is the basis for this. The Bons Sons Festival project is, in its essence, a catalyst for social interaction. According to the interviewee, the underlying idea "is not to go and work with the community. The community decided to make Bons Sons. It's very different, this changes the whole idea of the festival". Indeed, such an approach makes everyone responsible cells of the project.

5.3 Use of ICT

Finally, we investigated the role of technology in developing the Bons Sons Festival and inquired the interviewee about the tools, features, and platforms adopted. We found out that Digital technology is important in promoting Bons Sons, especially when it comes to promotion tools. Information about the festival is shared via social networks and press releases, as defined by the communication coordination. However, the interviewee is concerned with the poor Internet connections in the village, and states "we have great difficulty using Wi-Fi, because we'd need to make an enormous investment which we can't do because there's no fibber optics in the village, and to get the Wi-Fi response we need, we'd have to make an investment of 5 thousand euros". And he adds: "We still can't get generalized use of technology during the Festival, because there is no investment. We've approached some operators, but can't get that support and it's very expensive". This indicates that a better Internet connection could extend the results of the festival.

The interview revealed that the Bons Sons' team have the competences and autonomy to use ICT and digital technology. However, according to the director of Bons Sons, "folks aged 70–80 don't have such competences, nor the interest. Those who are about 60 years old use Facebook, share posts, and are thrilled by it. But it's a very conditioned, very limited, very targeted use". This lack of competence may hinder the further development of mediated networks with some of the village inhabitants.

Bons Sons uses a number of digital tools to disseminate the festival. These include the festival's webpage and mobile app as well and many other online tools. The festival website and mobile app display information about the origins the festival, the artists of the line-up, the purpose of each stage, and the facilities. They also show visitors how to get there, where to stay, and which services are available in the premises. The website also allows people to buy tickets and to recall previous editions of the festival, through photos and videos. Facebook and YouTube furthermore perform the role of building and developing social networks around the festival. The importance of these tools is recognised by the interviewee who says that, "for example, if there was no Facebook, there would never be Bons Sons". Finally, the interviewee also highlights the use of other online platforms, such as "clouds to store information, for timetabling, for virtual meetings" and states "we don't even use Skype anymore, we use lighter communication links over the Internet".

Regardless of the platforms and technology used, the festival's strategy prioritizes language. In the end, knowing how to publicize the festival on social networks implies using a language accessible to all users, especially those belonging to Bons Sons's target audience. Here, the interviewee mentions that the group is conscious of and concerned about the right way to reach their public. "In our communications we are careful to use simple, everyday language, so that everyone can understand what we are saying and feel involved", the interviewee comments.

Conscious of the importance of the user, responsible for spreading content, the interviewee adds, "The devices are not ours. Only the site is ours, in fact. Action on the networks, the algorithms that put content in one place or another, the payment that is now essential for the videos to have presence, that whole strategy depends on others". These concerns lead us to the discussion of the essence of dissemination and of governability via social networks. By publishing content on a social network one is waiving control over it. This is discussed by [39], who in a re-reading of the concept of governability of Michel Foucault's – where there is control of processes [40] – emphasizes the implicit lack of control present in social networks in digital environments. The results obtained with the study conducted with Bons Sons Festival show that the festival is aware of this phenomena and learning to adjust to it.

6   Discussion and Future Work

The way in which the festival articulates itself with the region challenges how one looks at rural areas; this makes it an innovative social initiative. In order to provide a musical experience, some festivals take over a deserted area for a few days to then abandon it and leave it behind once the event is over. Unlike these festivals, Bons Sons is an example of a mutually beneficial experience, with a clear learning experience dimension, to both visitors and the village. While attending the music festival, visitors are also offered the opportunity to engage with the landscapes, the people, the houses, the traditions, and the village as a whole. This creates a rich and meaningful social, cultural, touristic, and artistic ecosystem experience in which visitors can immerse themselves for the duration of their stay in Cem Soldos. To provide visitors with an opportunity to live a truly immersive experience, taking place in a 'real' village, with 'real' people, rituals, and places, the village had to re-organize itself and learn how to welcome and contribute to the event. In this way, Bons Sons is also a learning ecosystem for visitors and the village community. The organizers of the festival have managed to enthuse the whole community of the village of Cem Soldos to engage in the festival preparation and by doing so it has transformed a previously lethargic village into a thriving and vibrant region that people want to visit and where both older and younger generation want to live in.

The case of Bons Sons shows initiatives like this one offer great potential in the development of sustainable cultural and touristic projects. The analysis of the interview allowed us to gather a series of strategies, which are important for the development of an innovative project. This stresses the need for the development of policies which are specific to areas of low-density demographic areas, an idea that is also proposed by Nakala, Franque, & Ramos, [41, p. 155].

It is also a challenge to develop policies that promote development of territories since it would be necessary to avoid the "one size fits all" approach but taking into account particular contexts to allow people not to leave their places of residence and/or work in search of educational opportunities, which are often found in the big cities […].

The interview conducted with the director of Bons Sons also allowed us to observe that technologies deepen the sense of belonging, participation, and collaboration among community members, all of these essential in the development and continuity of a community [42, p. 247].

To correctly interpret the conclusions of this study, it is important to underline that those are drawn upon one single interview, with its main mentor, who is also the artistic director and organizer of the festival, and the interpretation of that interview by the authors. Arguably, carrying out interviews with other members of the Bons Sons team and of the village would have provided different points of view and detailed information on specific topics and the ability to triangulate data. Likewise, interviews with people who have attended and experienced Bons Sons would allow for a richer understanding of this case study. These are planned to take place in the near future.

In the future more studies on regional innovation initiatives are needed that specifically look into the role of technology mediation and identify key criteria and best practices for the use of technology in contexts of regional innovation. This will further our understanding on the subject and will allow for the comparison of different initiatives.

7 Conclusions and Lessons Learned

The interview with the artistic director of Bons Sons allowed us to gain valuable insight into how the festival has leveraged on the human and territorial potential of the village to develop the region where it takes place. This initiative provides an example of what the 'village of today' may be like, where rural communities can further develop their economic, cultural, and organizational autonomy and, therefore, add new dimensions to the regional ecosystem.

When analysing the organizational dimension of the project, we concluded that: (i) the definition and handover of strategic goals is clear and performed collectively; (ii) there is a discourse that all believe and take as their own, (iii) leadership is effective and aims at fostering motivation and promoting the growth of project, (iv) strategic goals and responsibilities are shared, (v) work is made by volunteers, whose diversity and engagement is fostered, and (vi) human resources are valued and respected.

Regarding the mediation dynamics, the following strategies were identified: (i) the festival leverages on the artistic and touristic potential and on the endogenous resources of the region to offer a high quality cultural and touristic experience; (ii) the tight relationships, cohesion, and synergies of the community network are fostered; (iii) the development of an empathy towards the audience, musicians, and village inhabitants, (iv) the collaborative knowledge development through networking and the engagement of participants and external partners; (v) the annual critical reflexion session about the developed activities and the collaborative resolution of any issues found; (vi) the empowerment of local communities through training and education, and (vii) the identification of points of improvement through reflection sessions with the community.

Finally, concerning the role of digital technologies, we observe: (i) the importance of the role of social networks to promote the project; (ii) the use of ICT solutions to share information about the festival and promote it at a national and international level; (iii) the use of digital technologies and online platforms as a means to facilitate communication, connection, and collaboration among team members; (iv) the development of a mobile app to promote and share information about the festival; (v) the use of simple and accessible language and an intuitive user interface that accommodates for all types of users, making the information about the festival available even to those who are digitally illiterate; (vi) the use of digital technology not only for communication, but also for digital inclusion, especially thinking about elderly citizens.

When looking at the crossover of the interview findings with TRI and DigComp we observe that the Bons Sons team is fairly optimistic and competent in using digital technologies. The team also appears to enjoy experimenting with new tools and keeps up to date with new trends, as the launching of the Bons Sons app confirms. Furthermore, the team observes the existence of alternative tools, using a lighter online solution as an alternative to Skype. While these examples indicate that the team is comfortable with the use of technologies, concerns about security, trust, and loss of control are raised regarding the information that is shared online among team members and with the community at large. Regarding digital competences in the areas of 'Information' and 'Communication' the analysis of the interview indicates that the Bons Sons team possess strong well-developed digital competences, and autonomy in its use. However, there are different levels of proficiency both within the core organization team and the community at large, where there is a concern with the autonomy of older populations.

This research also aimed to understand the extent to which the semi-structured interview developed for this study was a useful research instrument. After using it as a guide in the analysis of the interview conducted with Bons Sons, we confirmed this was an effective research tool to investigate the subject. The fact that it allowed us to uncover mediation dynamics, community networks, and the role of ICT in regional innovation makes it valid to apply in studies going forward.

Acknowledgments. We are grateful for the generosity of Luís Ferreira, the artistic director of Bons Sons, in making himself available to participate in the interview, which constitutes the basis for this study. This paper was developed under the support of the Research Program "CeNTER - Community-led Territorial Innovation" (CENTRO-01-0145-FEDER-000002), funded by Programa Operacional Regional do Centro (CENTRO 2020), PT2020.

References

1. Stuckey, B., Arkell, R.: Development of an e-learning knowledge sharing model. In: Australian Flexible Learning Framework Supporting e-Learning Opportunities. Australian Government. Department of Education, Science and training (2006)
2. Gunawardena, C., Hermans, M., Sanchez, D., Richamond, C., Bohley, M., Tuttle, R.: The theoretical framework for building online communities of practice with social networking tools. Educ. Media Int. **46**(1), 3–16 (2009)
3. Dabbagh, N., Reo, R.: Back to the future: tracing the roots and learning affordances of social software. In: Lee, M., McLoughlin, C. (eds.) Web 2.0- Based E-Learning: Applying Social Informatics for Tertiary Teaching, pp. 1–20. Information Science Reference, Hershey (2010)
4. Castro, E., Marques, T., Santinha, G. (coord.): Cidades Inteligentes, Governação Territorial e Tecnologias de Informação e Comunicação. Série Política de Cidades–2, POLIS XXI, DGOTDU, Lisboa, Julho (2008)
5. Schumpeter, J.: Capitalismo, socialismo y democracia. T.I. Ediciones Folio, Barcelona (1996)
6. Goggin, G., Ling, R., Hjorth, L.: Introduction: "MustRead" mobile technology research: a field guide. In: Goggin, G., Ling, R., Hjorth, L. (eds.) Mobile Technologies: Critical Concepts in Media and Cultural Studies: Volume I From the Telephone to the Mobile: Communication, Coordination, and New Connections, pp. 1–16 (2016)
7. Levinson, P.: New New Media. Pinguim, New York (2012)
8. Manovich, L.: Instagram and Contemporary Image. Manovich.net, New York (2016)
9. Castells, M.: Redes de indignação e esperança: movimentos sociais na era da internet. Zahar Editores, São Paulo (2013)
10. Castells, M.: A sociedade em rede – 8ª edição. Paz e Terra, São Paulo (2005)
11. Martin-Barbero, J.: De los medios a las mediaciones - Comunicación, cultura y hegemonía. Gustavo Gili, Barcelona (1991)
12. Scolari, C.: Hipermediaciones. Gedisa, Barcelona (2009)
13. Jenkins, H., Ford, S., Green, J.: Spreadable Media. NYU Press, Nova Iorque (2013)
14. Sousa, T., Tymoshchuk, O., Almeida, A.M., Santos, P.: An inclusive and multistakeholder approach to promote linguistic skills in children with special needs: the role of an online sharing platform. In: Proceedings of the 7th International Conference on Software Development and Technologies for Enhancing Accessibility and Fighting Info-exclusion, pp. 170–174. ACM, December 2016
15. Wenger, E., McDermott, R., Snyder, W.: Cultivating communities of practice: a guide to managing knowledge. Harvard Business School Press, Boston (2002)
16. Andrade, A.: Comunidade de prática: estudo de caso. Centro de Recursos em Conhecimento CRC. Associação Empresarial de Portugal (2005). http://www.knetpt.com/docs/ComunidadesPratica.pdf

17. Wenger, E., White, N., Smith, J. D., Rowe, K.: Technology for communities, CEFRIO Book Chapter, 18 January 2005 (2005)
18. Wenger, E., White, N., Smith, J.D.: Digital habitats: Stewarding technology for communities. CPsquare, Portland (2009)
19. Davis, F.D.: Perceived usefulness, perceived ease of use, and user acceptance of information technology. MIS Q. **13**(3), 319–341 (1989)
20. Parasuraman, A., Colby, C.: Techno-ready Marketing: How and Why Your Customers Adopt Technology. The Free Press, New York (2001)
21. Parasuraman, A.: Technology readiness index (TRI): a multiple-item scale to measure readiness to embrace new technologies. J. Serv. Res. **2**(4), 307–320 (2000)
22. Parasuraman, A., Colby, C.L.: An updated and streamlined technology readiness index : TRI 2. 0. J. Serv. Res. **18**(1), 59–74 (2015)
23. Eiamkanchanalai, S., Assarut, N.: Service quality and satisfaction of traditional and technology-enhanced services. In: Academy of Marketing Science Annual Conference, pp. 303–315. Springer, Cham (2017)
24. Koivisto, K., Makkonen, M., Frank, L., Riekkinen, J.: Extending the technology acceptance model with personal innovativeness and technology readiness: a comparison of three models. In: BLED 2016: Proceedings of the 29th Bled eConference "Digital Economy" (2016). ISBN 978-961-232-287-8
25. Meng, J., Elliott, K., Hall, M.: Technology Readiness Index (TRI): Assessing Cross-Cultural Validity. J. Int. Consum. Market. **22**(1), 19–31 (2009). https://doi.org/10.1080/08961530902844915
26. Pires, P.J., Alves da Costa Filho, B.: Fatores do índice de prontidão à tecnologia (TRI) como elementos diferenciadores entre usuários e não usuários de internet banking e como antecedentes do modelo de aceitação de tecnologia (TAM). RAC-Revista de Administração Contemporânea **12**(2) (2008)
27. Carretero, S., Vuorikari, R., Punie, Y.: The digital competence framework for citizens with eight proficiency levels and examples of use. https://doi.org/10.2760/38842
28. Lucas, M., Moreira, A.: DigComp 2.1: quadro europeu de competência digital para cidadãos: com oito níveis de proficiência e exemplos de uso. UA, Aveiro (2017). www.erte.dge.mec.pt/sites/default/files/Recursos/Estudos/digcomp2.1.pdf
29. Ferrari, A., Brečko, B.N., Punie, Y.: DIGCOMP: a Framework for Developing and. Understanding Digital Competence in Europe. eLearning Papers, **38**, 1–15 (2014)
30. Kotsanis, Y.: Models of competences for the real and digital world. Handbook of Research on Educational Design and Cloud Computing in Modern Classroom Settings, vol. 52 (2017)
31. Siiman, L.A., Mäeots, M., Pedaste, M.: A review of interactive computer-based tasks in large-scale studies: can they guide the development of an instrument to assess students' digital competence? In: International Computer Assisted Assessment Conference, pp. 148–158. Springer, Cham (2016)
32. Siiman, L.A., Mäeots, M., Pedaste, M., Simons, R.J., Leijen, Ä., Rannikmäe, M., Timm, M.: An instrument for measuring students' perceived digital competence according to the DIGCOMP framework. In: International Conference on Learning and Collaboration Technologies, pp. 233–244. Springer, Cham (2016)
33. Bryman, A.: *Métodos de investigação social.* OUP, Oxford (2012)
34. Bortolaso, I.V., Verschoore, J.R., Antunes Jr., J.A.: Práticas de gestão de redes de cooperação horizontais: proposição de um modelo de análise. Contabilidade, Gestão e Governança **16**(3) (2013)

35. Verschoore, J.R., Balestrin, A.: Fatores relevantes para o estabelecimento de redes de cooperação entre empresas do Rio Grande do Sul. RAC-Revista de Administração Contemporânea **12**(4) (2008)
36. Petter, R.R.H., Resende, L.M., Júnior, P.P.A.: Redes de cooperação horizontais e seus níveis de competitividade. RACE-Revista de Administração, Contabilidade e Economia **11**(2), 351–380 (2015)
37. Graells, P.M.: Las TIC y sus aportaciones a la sociedad. Departamento de pedagogía aplicada, facultad (2000)
38. Antikainen, M.: Facilitating customer involvement in collaborative online innovation communities, vol. 760, 94 p. VTT Publications, Espoo (2011)
39. Renó, D.: Narrativa transmedia y la des-gobernabilidad periodistica. Comunicação & Sociedade **34**(2), 141–161 (2013)
40. Foucault, M.: Microfísica do poder. Graal, Rio de Janeiro (1979)
41. Nakala, L., Franque, A., Ramos, F.: Public policies for quality assurance in distance learning towards territory development. In: Conference on Smart Learning Ecosystems and Regional Development, pp. 150–157. Springer, Cham (2017)
42. Andrew, N., Tolson, D., Ferguson, D.: Building on wenger: communities of practice in nursing. Nurse Educ. Today **28**(2), 246–252 (2008)

Sharing Community Data: Platform Collectivism for Managing Water Quality

John M. Carroll[(✉)] and Jordan Beck

The Pennsylvania State University, University Park, PA 16802, USA
`jmcarroll@ist.psu.edu`

Abstract. We are investigating community data, which is data gathered, analyzed, interpreted, and used by members of a local community. We discuss the ways that developing data literacy could result in more substantive civic participation and decision-making. Community members are already engaged in community data practices. Our goal is to help make these data more visible and accessible throughout the community and to engage the community at large in deliberation and planning with respect to its data. We report high-level outcomes of a community data hackathon that we facilitated along with a group of local water quality stakeholders. In order to design and develop an open community data platform and to build social infrastructure to sustain a long-term community data project, we plan to organize a series of these community-wide hackathon events.

Keywords: Community data · Data literacy · Hackathons
Platform collectivism

1 Introduction

Digital platforms mediate many activities of daily life. Early examples, like Parkatmyhouse, Airbnb, Craigslist, and Lending Club, were originally theorized optimistically as enabling a collectivist/sharing *re*conception of ownership (Botsman and Rogers 2011). However, the actual social consequences of digital platforms such as these are more varied and often more complex. Like other technologies, platforms that mediate collaborative interactions and exchange reorganize social structures and redistribute power. People can be harmed as well as facilitated by such infrastructures (Raval and Dourish 2016; Srnicek and Wiley 2016).

We draw inspiration from Ostrom's (1996) concepts of the commons and the coproduction of social goods. We describe the codesign of a community data platform by a collection of community stakeholders, including: local government, environmentalists, the local library, and community activists. Community data is the hyperlocal data that describes a community, is gathered and analyzed by community members, and is used by the community to guide deliberation and decision-making. For example, stakeholders interested in water quality may intend a dataset they collect to be useful to community leaders and citizens with regard to decision-making as it pertains to water conservation and land development issues.

© Springer International Publishing AG, part of Springer Nature 2019
H. Knoche et al. (Eds.): SLERD 2018, SIST 95, pp. 62–69, 2019.
https://doi.org/10.1007/978-3-319-92022-1_6

A weakness of contemporary democracy is the decline of participation and trust through disenfranchising social and technological infrastructures. Gurstein (2011), for instance, wrote about how an absence of efforts to equalize access and use of data will further increase social divides – especially with regard to marginalized communities. Platform capitalism is one manifestation of this. Our paper frames a remedial program of citizen engagement and community codesign motivated and focused by community data.

We envision a community data platform as an infrastructure for collating and discussing community data and its uses throughout the community. It addresses the direct concern of collective community resources being invisible and inaccessible to the community. For example, it enables the coproduction of environmental stewardship by a broad range of community stakeholders. Open datasets and educational outreach initiatives make certain kinds of community data available and accessible to community members. On the other hand, the data platform also raises issues of ownership, control, and responsibility in the local community context. Who owns the data? Who is responsible for its maintenance and curation?

In this paper, we discuss diverse local initiatives for water quality testing and threat mitigation that have arisen in the context of national and regional management (and lack of management). And we describe initial steps we have taken in a codesign process of bringing diverse stakeholders together to discuss community-level data sharing and management and of building a consensus for community action. We reflect on this work as an illustration of local codesign in which participants are ipso facto stakeholders, whose legitimate stakes and power are strengthened through greater visibility and engagement with other local stakeholders.

2   Community Data

In contemporary society, community membership is not a static state, but a set of dynamic tensions. Common interests such as shared territory and other resources, common purposes, such as collective security and child rearing, and common belief commitments can create the possibility of community. But community is embodied through the things people do, including social support, participation in collaborative projects, and identity work such as articulating the values that guide community decision making (Carroll 2012).

Through community data, a community describes itself to itself in order to understand its past, regulate its present, and plan its future. For example, several local organizations in our community collect water quality data with the goal of supporting community leaders and citizens in making sustainable decisions about water use and land development. However, it is a known issue within the open data and open government literature that availability of and access to data is insufficient to produce pervasive community participation (Janssen et al. 2012). Additional steps must be taken to ensure that community members make use of accessible data.

Community data is embedded in places, activities, and resources of the community. But data can also be represented and presented in various ways to make meanings and connections more accessible. Our goal is to identify needs and opportunities for citizens to better understand and contribute to data-driven civic participation as a focus for community innovation. We are interested in the possibility of leveraging data hackathons as means to engage citizens with community data.

Citizens of the past needed to possess basic levels of textual literacy. More recently, it has been argued that students ought to develop data literacy (Koltay 2015). In our view, data literacy is crucial for today's citizens. Citizens need to have facility in understanding and using data, and they need to be able to think critically and creatively about the many uses of data they might encounter. They will likely encounter a lot. From 2005–2010 the amount of data produced in the world increased tenfold. And, by 2020, it has been projected that the amount of data in the world might exceed 40,000 exebytes (Uhl and Gollenia 2014).

In our view, data literacy will help citizens identify relevant data, analyze and interpret data, participate more actively as community members, and use data in everyday civic contexts, such as public hearings about rezoning and land development and casual conversations with colleagues and neighbors. The use of data and data-driven argumentation to shape decisions and policies at the local level provides one opportunity to strengthen local democracy and democratic practices in general.

Citizens and local government may be committed to limited self-interest objectives, even though they pursue the same goal of an improved community (Sorensen 1977). Sorensen explained the problem as one where each party firmly believes that *their* approach is in everyone's best interest, and, gradually, power becomes the dominant factor for community actions. This leads to distrust and conflict between citizens and local government, which has been characterized as an integration gap. It is important to recognize this misalignment, identify possible underlying reasons, and better integrate citizens and the local government in community data practices.

Current information infrastructures, digital devices and sensors can empower citizens to initiate data-driven investigations of community concerns, such as: heritage protection, water quality management, energy regulation, and public by-way safety. Citizens can identify pertinent issues and research questions, coordinate with other citizens, gather and publish data sets online, and moderate community discussions and deliberations. Their role would be akin to those of citizen scientists with the exception that there would not necessarily be oversight from a subject-matter expert. Citizens would have greater autonomy and agency with regard to such projects. We want to help to make such initiatives easier to organize and carry out and more visible to the larger community.

These threads converge in the transformative possibility that data-enabled citizens could more constructively and effectively participate in and shape local governance that is itself data-oriented. Local governance has often failed through conflicts grounded in irreconcilable judgment and self-interest (Coleman 1957). Since there are other factors driving decision-making, data may not be a panacea for human conflict. Moreover, it may not be possible to realize a discourse that is free from judgment and interests. However, in our view, data can create an opportunity for more reasoned, rational discourse to come into being. At the very least, data are a shared community resource that has been under leveraged.

Our community partners are a collection of local organizations and nonprofit groups who monitor the quality of watershed resources. This work involves sustained and careful measurement of a range of properties including volume and velocity of water flow, temperature, suspended sediments, pH, coliform, lead, chlorine, nitrates, phosphates, and dissolved oxygen, among others. The groups operate in a loosely coordinated network, collecting and testing water samples, organizing data sets and document collections, and sharing data with local and state government and with community members and groups.

In addition, they carry out various water quality protection/mitigation activities (such as creating riparian buffers in the streams of the local watershed), and organize and participate in local meetings, workshops, and other initiatives aimed at informing community members of the state of the watershed, enumerating long and short term threats to the watershed, and proposing courses of preservative or corrective action. Even in the relatively small community we are studying (about 150 square miles with a population of 93,000), water quality is a complex sociotechnical project.

As part of a long-term research project focused on community data, we are investigating the role of hyperlocal data in contemporary community. Data pertaining to a community and its locale, that is, data gathered, analyzed, interpreted, and used by members of a local community, is community data. There can be many different kinds of community data. Examples include but are not limited to: water and air quality, demographics, narratives about historically significant places, photographs of weather events, and so forth.

With our partner organizations and groups, we are exploring the concept of a *community data platform* - a digital commons for community data, including water quality data. The data platform would make community data visible to the community. It would integrate dispersed sources and types of data, and facilitate more active engagement by a wider range of community members, for example, supporting discussions of data and interpretations of data, as well as proposals for further data-oriented projects. In order to design and develop the community data platform as well as build social infrastructure that could contribute to the long-term sustainability of such a resource, we aim to leverage best practices in intensive, one-day hackathon events.

3   Water Quality Hackathons

What could public hearings look like if more citizens were able to identify data relevant to a pressing civic issue and accurately analyze and interpret it? We believe that their contribution would almost certainly be more substantive and convincing. Consider the following scenario:

> *A recent proposal to irrigate local soccer fields with beneficial reuse water caught Hamid's attention. Beneficial reuse water is wastewater that has been processed to meet state and federal regulations for cleanliness, but Hamid can't get past the fact that his kids might play soccer on a field that has been irrigated with "toilet water." Using the data platform, Hamid learns about beneficial reuse, and analyzes data describing beneficial reuse water before and after treatment. He sees key differences in hardness, conductivity, and dissolved oxygen levels, and compares these values with official state regulations – noting that the values are well within safe ranges.*

We envision that Hamid's contribution to a public hearing about whether to make use of beneficial reuse water to irrigate soccer fields would be more reasonable if not more compelling following their analysis and interpretation of data via some kind of open community data platform. Toward this end, we want to facilitate a series of community data hackathons: workshops where multiple stakeholders and subject matter experts generate ideas, plans, designs, or prototypes in response to a design brief that we will provide (Morelli et al. 2017). Hackathons and other kinds of community workshops have been used to engage citizens around open data and pertinent civic issues.

A key element of our approach is that participants (community members) contribute ideas and proposals directed at possible future courses of action. They do not merely generate diverse ideas (as in brainstorming) or criticize existing approaches. For example, at an early-stage hackathon, it might be possible to articulate different levels of data literacy, identify learning objectives to achieve those literacy levels, and plan a series of community workshops aimed at achieving those objectives. This means that our participants would not only be charged with advancing planned work but in planning the work as well, including (potentially) defining the metrics for its success. We see this level of citizen participation as cultivating a sense of community ownership of the project, which could be a crucial step towards *sustaining* a long-term community data project.

In November 2017, we organized a small community workshop on water data. Three of the key community groups involved in local water quality monitoring participated, along with a member of the local government. The groups were all very eager to share information about their practices and experiences. Beyond our initial remarks we made the decision to remain open to discussing topics of interest to the participants. Over the course of an hour and a half, we gained valuable insights about database management problems, coordination and cooperation (or lack thereof) between groups, as well as pain points.

Interestingly, although they all knew one another, and recognized common interests and practices, they did not seem aware in detail of what other groups were doing. For example, when one group described the quality assurance processes they follow in order to verify the accuracy of their water quality data, other participants began to compare and contrast this description with their own quality assurance practices. We raised the notion of integrating data sets to produce a more comprehensive picture of local water quality. The reaction of the group was positive, but it was also clear this had not been considered before. However, it was striking that members of the workshop were (and are) working on an online resource that describes aspects of the local watershed system (http://www.springcreekwatershedatlas.org).

The vision of their Watershed Atlas is to provide a vivid and accessible online resource for public use and engagement. This suggested a key contrast to us: Our partners are well aware that the watershed is a community commons, and that descriptions of the watershed can make this more visible and useful to the community.

However, the idea that their multiple datasets could be programmatically aggregated into a dynamic data resource for the community is a distinct step beyond that. In the months since this original hackathon experience, we have had a series of smaller discussions with our partners about data practices and about the concept of a community data platform. Our plan is to organize further hackathons to explore concepts for sharing and supporting data and data tools with the community with the core water data community and with other constituencies throughout the community. In some cases, we plan to use design probes as a way to invite community members into a local codesign process.

The stakeholders at our inaugural hackathon, for instance, provided some high-level design ideas for an integrated platform in the form of questions and comments about possible needs and requirements. We see our community partners as collaborators and co-investigators. *Collaborators* help formulate the designs of projects, and *co-investigators* share equal responsibilities with scientists in projects (Bonney et al. 2009).

We want to leverage our experience in community-scale participatory design (Carroll et al. 2000; Carroll 2012; Carroll and Rosson 2013) to engage a wide range of community stakeholders in large-scale action research. Participatory design is both an inclusive and equitable approach to design and a method for engaging with and learning about stakeholder values, practices and knowledge (Béguin 2003; Simonsen and Hertzum 2012).

4 Discussion

We are impressed that the existing water quality groups function simultaneously as a community service activity, a context for learning and practicing new technical skills, and a source of social support and interaction. Many of the participants are older adults creating second careers in environmental monitoring. The members of the groups seem eager to share their experiences and personally motivated to expand the model of their groups more generally throughout the community.

Our partners see it as inherently problematic that they each carefully manage a separate data view into community water, and that there is no way to integrate these views into a whole. For example, several of our partner organizations share datasets with one another; many containing, overlapping data points. Is this the most effective use of resources? Can the redundancy be leveraged to measure and visually convey data quality? On the other hand, perhaps these overlapping measures are complementary since they originate at different field sites. Taken together, multiple measures of pH or chloride levels could yield a more complete picture of a watershed. The extent to which this sharing/overlap is pervasive is unknown. However, that the practice of sharing exists could be evidence for the value of an integrated data platform.

Existing stakeholders clearly understand the value in community learning and participation with respect to water quality data. For example, one group is involved in creating an online overview of various water quality issues and is discussing the utility of integrating that with actual data, raw or curated. The regional public library has expressed an interest in hosting the community data platform. We frame this as *platform collectivism*: instead of specializing producers and consumers of platform-mediated services, participation is a reciprocal co-production (Carroll et al. 2016).

This seems to be a case where a modest technological scaffold could enable broad and diverse new social and civic possibilities. First, citizens could understand water quality issues as issues they can act upon and contribute to directly. They can learn about data, data analysis, and data-driven thinking in a real and concrete context that is meaningful to and consequential for them and their families.

The data platform would make community water monitoring activity more visible to its stakeholders throughout the community, including governance structures. It would codify incentives and provide a commons for enhancing the coherence of community water monitoring. In these senses, the platform would support collectivism in the community it supports.

Community data activities can be innovations in citizen participation. They are creative appropriations of technology to engage with, monitor, and manage local community. Some of the local groups we have talked to participate in networks and, occasionally, cooperate with state and local government agencies, but they are not hierarchically supervised. We have begun our work by taking steps to systematically understand how and why local water quality groups operate as they do, who they work with and how, and what constraints and limitations they face.

We are mapping the local stakeholders in water quality activity, including government, nonprofit organizations, and businesses, to identify opportunities for citizens to better understand and contribute to data-driven civic participation as a focus for community innovation. The local sociotechnical innovations of community data can be better understood, more coherently facilitated, and can be codified as national models where appropriate.

5   Conclusion

We suggest that community data can strengthen community itself in two ways. First, this hyperlocal data codifies the common interests that enable community. To the extent that members can be motivated to feel belonging and connection, to engage and participate with one another, or to exchange social support, appreciating their own shared interests and the possibility of developing and advancing those shared interests, should evoke and strengthen sense of community. Neolithic communities were shaped by the management of their grains and other farming products. Modern communities may similarly take shape from effectively gathering, analyzing, and using their data.

We cannot directly investigate the future of citizenship and the role of data literacy and data-driven thinking in citizenship. But it is not too soon to investigate how community members are already appropriating community data practices, and to work with them to facilitate these practices, to make them more visible to the larger community, and to help them reflect on new affordances and meanings of these practices. Data would be a crucial part of how a community presents and projects itself to higher levels of government – characterizing and defending its interests in larger-scale policy and initiatives. Such a role for data would make data literacy fundamental to citizenship.

Acknowledgements. This work is supported by a Faculty Fellowship from Student Engagement Network of Pennsylvania State University.

References

Béguin, P.: Design as a mutual learning process between users and designers. Interact. Comput. **15**(5), 709–730 (2003)

Bonney, R., Ballard, H., Jordan, R., McCallie, E., Phillips, T., Shirk, J., Wilderman, C.C.: Public participation in scientific research: defining the field and assessing its potential for informal science education–a CAISE Inquiry Group Report. Center for Advancement of Informal Science Education, Washington, DC (2009a)

Botsman, R., Rogers, R.: What's mine is yours: how collaborative consumption is changing the way we live (2011)

Carroll, J.M.: The Neighborhood in the Internet: Design Research Projects in Community Informatics. Routledge, New York/London (2012)

Carroll, J.M., Rosson, M.B.: Wild at home: the neighborhood as a living laboratory for HCI. ACM Trans. Comput. Hum. Interact. **20**(3) (2013). Article 16

Carroll, J.M., Chen, J., Yuan, C.W., Hanrahan, B.V.: In search of coproduction: smart services as reciprocal activities. IEEE Comput. **49**(7), 26–32 (2016)

Carroll, J.M., Chin, G., Rosson, M.B., Neale, D.C.: The development of cooperation: five years of participatory design in the virtual school. In: Boyarski, D., Kellogg, W. (eds.) DIS 2000: Designing Interactive Systems, Brooklyn, New York, 17–19 August, pp. 239–251. Association for Computing Machinery, New York (2000)

Coleman, J.S.: Community Conflict. Free Press, Glencoe (1957)

Gurstein, M.: Empowering the empowered or effective data use for everyone? First Monday **6**(2) (2011)

Janssen, M., Charalabidis, Y., Zuiderwijk, A.: Benefits, adoption barriers, and myths of open data and open government. Inf. Syst. Manag. **29**, 258–268 (2012)

Koltay, T.: Data literacy: in search of a name and identity. J. Documentation **71**(2), 401–415 (2015)

Morelli, N., Aguilar, M., Concilio, G., Götzen, A.D., Mulder, I., Pedersen, J., Torntoft, L.K.: Framing design to support social innovation: The open4citizens project. Des. J. **20**(1), 3171–3184 (2017)

Ostrom, E.: Crossing the great divide: Co-production, synergy, and development. World Dev. **24**(6), 1073–1087 (1996)

Raval, N., Dourish, P.: Standing out from the crowd: emotional labor, body labor, and temporal labor in ridesharing. In: Proceedings of the 19th ACM Conference on Computer-Supported Cooperative Work & Social Computing. ACM, New York, pp. 97–107 (2016)

Simonsen, J., Hertzum, M.: Sustained participatory design: extending the iterative approach. Des. Issues **28**(3), Summer 2012

Sorensen, D.C.: Reverse migration and the rural community development problem. West. J. Agric. Econ. **1**(1), 49–55 (1977)

Srnicek, N.: Platform Capitalism. Wiley (2016)

Uhl, A., Gollenia, L.-A.: Digital enterprise transformation: a business-driven approach to leveraging innovative IT. Routledge, New York (2014)

Tiles-Reflection: Designing for Reflective Learning and Change Behaviour in the Smart City

Francesco Gianni[✉], Lisa Klecha, and Monica Divitini

Department of Computer Science,
Norwegian University of Science and Technology, Trondheim, Norway
{francesco.gianni,monica.divitini}@ntnu.no, lisark@stud.ntnu.no

Abstract. Modern cities are increasing in geographical size, population and number. While this development ascribes cities an important function, it also entails various challenges. Efficient urban mobility, energy saving, waste reduction and increased citizen participation in public life are some of the pressing challenges recognized by the United Nations. Retaining livable cities necessitates a change in behaviour in the citizens, promoting sustainability and seeking an increase in the quality of life. Technology possesses the capabilities of mediating behaviour change. A review of existing works highlighted a rather unilateral utilization of technology, mostly consisting of mobile devices, employment of persuasive strategies for guiding behaviour change, and late end-user involvement in the design of the application, primarily for testing purposes. These findings leave the door open to unexplored research approaches, including opportunities stemming from the Internet of Things, reflective learning as behaviour change strategy, and active involvement of end-users in the design and development process. We present Tiles-Reflection, an extension of the Tiles toolkit, a card-based ideation toolkit for the Internet of Things. The extension comprises components for reflective learning, allowing thus non-expert end-users to co-create behaviour change applications. The results of the evaluation suggest that the tool was perceived as useful by participants, fostering reflection on different aspects connected to societal challenges in the smart city. Furthermore, application ideas developed by the users successfully implemented the reflective learning model adopted.

Keywords: IoT · Reflective learning · Smart cities

1 Introduction

Nowadays, more than half of the world's population lives in cities [6]. Urbanization at this scale ascribes cities a key function, as cities have a vast influence on economic and social aspects, as well as environmental impacts [1]. However, as cities grow, so do the challenges they face. Challenges comprise, among others,

© Springer International Publishing AG, part of Springer Nature 2019
H. Knoche et al. (Eds.): SLERD 2018, SIST 95, pp. 70–82, 2019.
https://doi.org/10.1007/978-3-319-92022-1_7

a difficulty in waste management, scarcity of resources, air pollution, human health concerns, traffic congestion, and deteriorating infrastructures together with increasingly complex social problems [11]. Those issues exert a harmful influence on habitability, and measures urgently need to be taken to ensure sustainable conditions. In this context the notion of smart city has increasingly gained in notoriety, describing cities that devise smarter ways to manage the challenges imposed on them [5]. Retaining livable cities, and achieving urban sustainability goals requires a change in behaviour towards more sustainable societies [15, 18].

While information and communication technologies (ICTs) and the Internet of Things (IoT) appear to be the common denominator in defining a smart city [12], it is increasingly recognized that a smart city is indeed a multidimensional and multifaceted concept and, therefore, smart cities should be studied and analyzed on the basis of multiple components [10]. In Nam and Pardo's conception [22], a pervasive IT infrastructure is essential, but not enough without the engagement and collaboration between city stakeholders. Hence, also human factors are stressed, emphasizing such things as creativity, education and social inclusion. Smart people is a concept that is crucial, as well as smart communities, underlining that collective intelligence and social learning make a city smarter [22].

Technology can help also in mediating behaviour change. The work presented in this paper aims to facilitate end-user participation in the design of behaviour change applications for cities, which utilize IoT as mediating technology, and reflective learning as behaviour change strategy. This is motivated by a systematic review of urban mobility behaviour change applications, which highlighted several opportunities in areas of technology usage, behaviour change strategies, and end-user participation in the design and development process of such applications [16].

While consumers voice concerns about the impact of their behaviour on the environment or on the society, their actions do not conform with those worries [2]. This gap between pro-environmental values and pro-environmental behaviours can be partly explained by the notion of routines or habits, being behaviours that are highly automated, requiring little cognitive effort to be performed [19]. With almost a half of everyday activities being classified as habitual [30], finding effective measures and strategies to break and replace those habits with more sustainable ones is crucial and challenging.

We chose to support this process by creating Tiles-Reflection, an extension for the Tiles toolkit, a card-based ideation toolkit for IoT applications [21]. The extension centres on behaviour change applications in the context of a smart city, with reflective learning posing as the utilized strategy to foster behaviour change. The feasibility and utility of this approach are assessed in workshops with citizens. Reflective learning was chosen as an approach to promote slowness in learning and understanding, meaning to provide people with time for reflective activities and conscious use of technology [13]. Similarly, the concept of *slow change* interaction design, evolves around the idea of creating technologies that facilitate attitudinal and behavioural change over time [29]. Slow change acknowledges that change may take a long time, being an endless and difficult process that should not be forced on people, essentially requiring people to take the first step [29]. Recalling identified issues of persuasion [16], reflection may prove as an alternative strategy for behaviour change. Reflection has been described as having a strong social dimension and being often accomplished collaboratively [17]. This characteristic might be particularly beneficial in a smart city context, in which it should be considered that citizens not solely constitute individuals, but also communities and groups [5].

Boud et al. [3, p.18] ascribed reflection particular significance in any form of learning. They see reflection as "a form of response of the learner to experience", in which the experience is recalled, pondered on, and evaluated in order to gain new understandings and appreciations. In short, reflection turns experience into learning. The trigger for reflection may thereby emerge from an external event, or from a state of inner discomfort, but likewise from more positive states, for instance upon the successful completion of a task. Krogstie et al. [17] follow this perception of reflective learning, they see it as the "conscious re-evaluation of experience for the purpose of guiding future behaviour". Boud et al.'s model of reflective learning further informed the development of Krogstie et al.'s Computer Supported Reflective Learning model, hereafter referred to as CSRL model [17]. The model is presented as a four-staged reflective learning cycle, comprising plan and do work (1), initiate reflection session (2), conduct reflection session (3), and apply outcome (4), each encompassing a number of activities. Results from these stages feed as input to the next, including data on work, a frame for reflection, the reflection outcome, and a change on the activity.

2   Related Work

In a previous work, we surveyed technological applications for behaviour change in the city [16]. A systematic literature review was conducted, exploring previously envisioned or implemented solutions, addressing urban mobility behaviour change. Three areas were thereby mostly of concern: (i) the utilized technology, (ii) the behaviour change strategies, and (iii) how end-users participated in the design and development of these applications. The review revealed an unilateral use of technology, favouring mobile applications. Persuasive strategies were foremost guiding behaviour change, and end-users appeared to be involved mainly for testing purposes, late in the process. Furthermore, it emerged that most of the applications were primarily tailored for individual use, rather than collectives. These conclusions exposed several opportunities in the aforementioned areas that are unexplored, briefly summarized in the following.

Adopted Technology - Ubiquitous technology and the IoT emerged as opportunities for breaking new technological grounds. This approach is supported by research on technology-enhanced smart city learning [8]. In a systematic mapping of the topic, mobile devices were likewise identified as the prevailing category, whereupon interactive objects and the IoT are recommended, as they provide novel interaction modalities. This argument was further reinforced in a subsequent article, in which the limited interaction possibilities of mobile devices were addressed [9].

Behaviour Change Methodology - Due to the identified shortcomings of persuasive systems, previous research advocates a shift from prescription to reflection [4]. Reflective learning resulted as a more efficient and long-lasting approach than strategies based on persuasion.

User Participation - Active citizen or end-users participation is emphasized as a prerequisite for behaviour change system development in a smart city. This view suggests a move from "making technology designers arbiters of all things sustainable" towards "a more deeply involvement of the users" [4]. On the same line, Pettersen and Boks [26] describe participatory design as a method that can "contribute to the development of socially robust, ethically justifiable technologies for behavioural change".

End-user involvement is advantageous also in the context of sustainable behaviour change applications. Participatory design approaches facilitate democratization of design, empower people, and not least, emphasize diversity in the groups they include [26]. However, user involvement in the development of behaviour change applications is scarce in the smart city domain [16,28], as well as in other research fields concerned with sustainability themes [4]. Only a small percentage of works report the adoption of participatory design methods, or other forms of user engagement, during the design phase of the applications.

3   Supporting Tools

We investigate the possibility to extend the Tiles toolkit [21], which has proved to be an appropriate tool for co-design of IoT applications targeting a diverse set of stakeholders, ensuring inclusive design and support creations that are citizen driven. However the tool does not specifically target the design of behaviour change that is rooted in reflective learning. Therefore in this paper we propose Tiles-Reflection, an extension of the original toolkit introducing the reflective learning component.

Tiles is a card based design toolkit for IoT, meant to serve as a source of guidance and inspiration when brainstorming applications involving augmented objects. For this purpose, it encompasses a set of 110 design cards and a workshop technique, structuring the use of the cards by means of a playbook reporting step by step instructions, and a cardboard.

Seven distinct decks of cards are devised to abstract the complexities of IoT technologies, making the concepts tangible for non-experts. Customizable personas and scenarios provide constraints to help focusing the idea generation process, while further facilitating participatory design of smart object applications as user-centred design artifacts. Another notable aspect of the toolkit is its flexible adaptability to a variety of domains [21]. Creative workshops can be conducted with 2–6 participants, and are meant to be supervised by professionals. The playbook and cardboard foremost guide the activities during the workshop. Within the playbook, seven design steps are described, walking the participants through the design of one or more augmented objects.

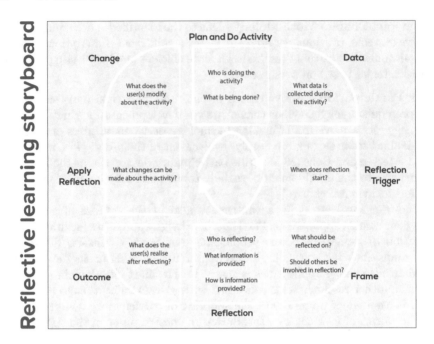

Fig. 1. The reflective learning storyboard used during the workshops.

The original Tiles toolkit has been extended to support the development of IoT applications promoting reflective learning. Four suitable stages were identified at which reflective learning had to be introduced, two of which were directly integrated into already existing Tiles components (*mission* and *criteria* cards), while two novel elements were introduced in the form of an additional reflective learning storyboard (Fig. 1) and a more detailed frame to describe the persona, based on a simplified version of the *persona in practice* model, introduced in the MIRROR design toolbox [25]. In addition, to facilitate the inclusion of multiple target users in the reflective idea, users were asked to list a possible set of persons or communities pertaining to the *social circle* of the chosen persona (Fig. 2).

The additional, reflection oriented, *misson* and *criteria* cards served to focus the design of the application from the very beginning and provided criteria to retrospectively assess, and eventually fix, the reflective dynamics embedded by the participants in the application idea. The *persona in practice* model provided additional structure in the definition of the target user. Participants were encouraged to define upfront the sub-optimal behavior to be changed through the reflective IoT application, and the attitude towards technology and behaviour change of the persona. The reflective learning storyboard (Fig. 1) essentially corresponds to the CSRL model [17] in shape and content, however, some aspects were altered to better fit the context of behaviour change, considering that the CSRL model has been designed to depict reflective learning in the workplace. The model is composed of four stages interleaved by four transitions. Starting from an initial activity,

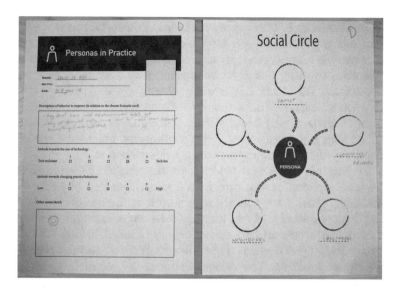

Fig. 2. Persona in practice and social circle models used during a workshop.

data is presented to the user in order to trigger the reflective process, the outcome of which fuels and describes a concrete change that is then applied to modify and improve the original activity, closing thus the reflective circle. Each stage is represented on the storyboard by a square, while the transitions represent the corners of the diagram.

4   Design Process and Evaluation

Three focal assessment objectives for the workshops were researched:

- **[O1] Usefulness** - the perceived usefulness of the Tiles toolkit and workshop in supporting the design of reflective applications;
- **[O2] Reflective Learning** - the effectiveness of Tiles-Reflection in supporting the design of reflective IoT applications;
- **[O3] Co-Design** - the perceived usefulness and intrinsic motivation provided by the co-design approach of the workshop, stimulating inclusion and participation.

Human-computer-interaction defines usefulness as the sum of usability and utility [23]. In order to assess the cards' usefulness, this equation was used to frame the evaluation. While usability describes the ease and pleasantness of using the tool, utility relates to whether it provides what the user needs [23].

Data about the workshops were collected in two ways. Prior and after the workshop the participants were asked to fill out questionnaires. Data collected included information about participants' demographics and their perception of the workshop experience. This approach yielded in quantitative data, which were aggregated and analyzed using a spreadsheet software. During the workshop, notes documenting the observed dynamics were taken and a camera recorded the cardboard, capturing how the participants interacted with each other and with the extended Tiles toolkit. Eventually, the video footage allowed to extract qualitative and quantitative data about: (i) issues with the Tiles elements (playbook, cards, cardboard); (ii) issues with provided guidance and information; (iii) suggestions about improvements; (iv) time spent on each step of the workshop and in total; (v) the devised augmented object application.

Lastly, group interviews were held to obtain a more thorough understanding of participants' opinion on the workshop matter. Research objectives O1 and O3 were thereby mostly assessed through questionnaire data, whereas research objective O2 was assessed solely with the gathered video and photo material. The evaluation was then informed by the guidelines outlined in [24]. The design of the reflective learning extension of the Tiles toolkit was performed and refined during multiple iterations. The evaluation focus is kept mainly on the Tiles-Reflection extension, since the generic workshop and toolkit have already been evaluated [21]. For each iteration, one or more workshops with the users were performed to evaluate the design and collect feedback. We hereby briefly present such iterations, which are summarized in Table 1.

I - two master students of the department of computer science were invited to test the workshop protocol and the Tiles extension prototype during a pilot workshop.

II - three researchers participated in the second iteration, the workshop protocol, cards and reflective storyboard experimented in the pilot were finalized and employed during this phase.

III - a rather diverse group of users took part in the third evaluation. The four groups included high school students, municipality employees, freelancers, entrepreneurs and programmers from a local coworking space. In an attempt to possibly reduce the time needed to browse the many cards, additional mission cards aside of the preset reflective learning mission card were removed.

IV - the last iteration comprised five groups composed by computer science university students. In order to decrease the level of support needed by the participants and to provide better guidance during the workshop, the *persona in practice* and *social circle* models were introduced in this iteration.

Table 1. Details of the workshops.

Iteration	N	N. groups	Age	Occupation
I	2	1	19–27	University students
II	3	1	40–55	Researchers
III	13	4	17–50	Students, municipality, entrepreneurs
IV	25	5	20–40	University students
TOT	43	11		

5 Results

We now present the results of the four iterative evaluations, following the research objectives previously presented in Sect. 3.

O1: Usefulness - Data from the questionnaires, presented in Fig. 3, shows that the workshop was perceived as useful for the design of reflective IoT applications. The following statements are reflected in the statistics:

- **S1:** The criteria cards helped me to evaluate my idea with respect to reflection;
- **S2:** It was easy to design an application that supports reflection;
- **S3:** The reflective learning storyboard was easy to use;
- **S4:** I can imagine conducting a workshop using Tiles without guidance, on my own or in a group;
- **S5:** I can think of other scenarios in the city where IoT could be used.

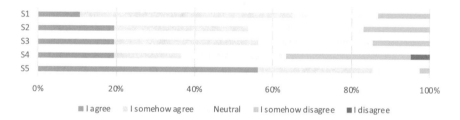

Fig. 3. Results of the questionnaire statements relative to O1.

The overall utility of the tool was rated positive (statements 1–3), as more than 50% of the participants adjudged the tool to be easy to understand and helpful during the ideation of reflective applications. Particularly well perceived was the statement pertaining the potential of the tool for other urban scenarios. Mixed opinions were extracted from S4, where there's near perfect balance between participants that can imagine conducting a workshop on their own and participants who can't.

O2: Reflective Learning - Most of the groups were successful in developing an idea involving IoT and reflective learning. Among the ideas created during the workshops, an augmented wheelchair that provides feedback when a more accessible route in the city is available; a mug which shows to people in the room random tweets posted by the owner, to increase privacy awareness about the information shared online; a smart recycle bin that reacts with emojis and visualizes environmental impact data about the trash when kids fail to recycle in the correct way.

We now analyze in more detail the smart recycle bin application. The cardboard, cards and reflective storyboard used during the workshop can be seen in Fig. 4. The chosen *persona in practice* is a group of children 4–7 years old, which have no education in social environmental habits. The application provides the children with information about the trash and a direct feedback whenever they use the bin. These outputs act as reflection triggers, envisioned during the second stage of the CSRL storyboard, and are intended to have a double effect on children's perception of environmental sustainability. On one side they are confronted with the impact of the trash produced, and in addition they receive a negative feedback, in form of sound or emoji, if the trash is not placed in the correct recycle bin. The intended outcome of the reflective learning process is an increased awareness of the environmental impact of waste and knowledge about how to correctly recycle trash. The devised application is mapped into the CSRL model's stages and transitions through the reflective learning storyboard reported in Fig. 1, to reveal the degree to which the idea is potentially able to support reflection.

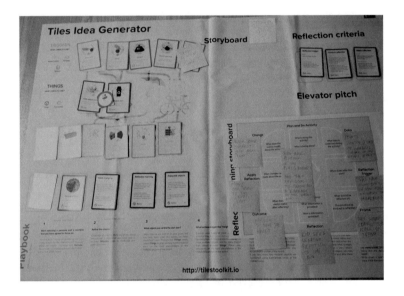

Fig. 4. The cardboard, cards and storyboard at the end of one of the workshops.

O3: Co-Design - The results collected through the questionnaires are shown in Fig. 5, the following statements are reflected in the statistics:

- **S6:** I think that being involved in the design of such applications would motivate me to later use them;
- **S7:** I think that involving citizens/end-users in the design of such applications will result in more innovative solutions.

The majority of participants strongly agreed with both the statements, underlining that the respondents considered the involvement of end-users in the design process as crucial.

6   Discussion

O1 - Observations and questionnaire results underpin that the Tiles-Reflection workshop was very well received, and perceived as useful by participants. Diverging opinions were only observed in relation to if people could imagine conducting a workshop by themselves, without guidance. The statement essentially aimed at assessing whether people may utilize the Tiles-Reflection workshop themselves for ideation, hence for citizen-driven, bottom-up innovations that would address their specific needs. This view is in line with the concept of "empowering people to devise ways to run their daily lives as smartly as possible, making their extended community –the actual embodiment of a city– smarter, too" [27]. Despite the questionnaire results on the matter, we observed an improvement in the ability of the groups to work more independently, while still delivering relevant ideas. This improvement has been registered during iteration *IV*, and might be due to the additional support provided by the extended persona models. For comparison, all the groups in the first three iterations were directly supported by at least one of the authors for the whole length of the workshop, while during iteration *IV*, only one of the authors supervised the workshop, attended by five groups simultaneously. Results suggest that the Tiles-Reflection workshop is indeed useful for co-creation practices. It is, therefore, more reasonable to consider the Tiles tool with the reflective learning extension, as a mean for stakeholders to provide meaningful input to the design of reflective learning applications, rather than a way for novices to design an application in all its details by themselves. Eventually, results of the different design and evaluation iterations further suggest that the format improved over time.

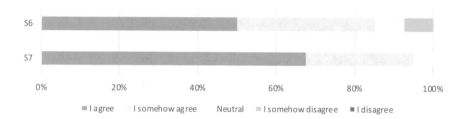

Fig. 5. Results of the questionnaire statements relative to O3.

O2 - It was pleasant to observe that participants fully immerse themselves in the task scenario, discussed personal experiences regarding sustainability, and put themselves in the position of future users. Overall, the outcome illustrate that all groups managed to incorporate reflection sufficiently in their application design. Some groups had more difficulties envisioning how technology could facilitate the last step of the reflective process, explaining how to modify the practices applying the results of the reflective process [7]. Krogstie et al. [17] state that certain aspects of the CSRL model are "more or less explicit", such as the reflection frame or reflection outcome, or "more or less elaborated" and "brief and closely integrated with other activities", such as the reflection trigger stage. Hence, researchers and designers eventually involved in the co-creation process, can help tailoring the reflection elements to the specific ideas envisioned by the participants, emphasizing the reflective steps when needed. Pertaining to the reflective learning additions of the Tiles toolkit, participants appeared to be confident in using the cards and the storyboard, with some groups even explicitly using the storyboard to pitch their idea to the authors. Finally, since many of the participants were new not only to the concept of reflective learning, but also to the nature and definition of an IoT application, it was particularly challenging for them to first get familiar with the new notions and then successfully apply them to create an innovative solution.

O3 - Participants in iteration *III* affirmed that the co-design workshop facilitated their reflection, they even voiced interest in using the workshop as a mean to reflect with their co-workers and management. Most participants perceived end-user involvement in the development of the reflective applications as beneficial. This viewpoint underlines people's interest in participatory design, which might lead to an increase of both innovation and adoption, thus research on co-creation tools appears particularly eligible.

Summarized, results from the workshops showed that an adequate time frame, illustrative examples regarding IoT and reflective learning, as well as participants actively communicating and collaborating during ideation, are all crucial factors for a successful workshop session. As anticipated by participatory design methodologies, workshop facilitators were confirmed as an essential component to guarantee a valuable outcome, their role in supporting the users during the ideation process has been once again fundamental.

7   Conclusions

With this work, we proposed a tool to involve citizens in the ideation of technological applications, promoting sustainability, behaviour change and lifelong learning through reflection. Shifting the focus to cities has gained in momentum, in particular enabling citizens to take an active role in the development of cities of the future [14]. Bottom-up innovation and collaboration are needed alongside top-down approaches [14]. Concepts such as Human Smart Cities [20], and the notion of Smart Citizen [14] underpin this standpoint. Tools like Tiles-Reflection can contribute to these objectives by providing means for ideation of reflective IoT applications with diverse stakeholders. Furthermore, as the Tiles toolkit provides possibilities for prototyping [21], citizens can become makers and active contributors. Eventually, "there can be no smart city without smart citizens" [27].

Acknowledgements. We would like to thank all the students and volunteers that participated in the evaluation workshops, and Dr. Simone Mora for his work on the Tiles toolkit.

References

1. Albino, V., Berardi, U., Dangelico, R.M.: Smart cities: definitions, dimensions, performance, and initiatives. J. Urban Technol. **22**(1), 3–21 (2015)
2. Bhamra, T., Lilley, D., Tang, T.: Design for sustainable behaviour: using products to change consumer behaviour. Des. J. **14**(4), 427–445 (2011)
3. Boud, D., Keogh, R., Walker, D.: Reflection: Turning Experience into Learning. Routledge (2013)
4. Brynjarsdottir, H., Håkansson, M., Pierce, J., Baumer, E., DiSalvo, C., Sengers, P.: Sustainably unpersuaded: how persuasion narrows our vision of sustainability. In: Proceedings of the SIGCHI Conference on Human Factors in Computing Systems, pp. 947–956. ACM (2012)
5. Chourabi, H., Nam, T., Walker, S., Gil-Garcia, J.R., Mellouli, S., Nahon, K., Pardo, T.A., Scholl, H.J.: Understanding smart cities: an integrative framework. In: 2012 45th Hawaii International Conference on System Science (HICSS), pp. 2289–2297. IEEE (2012)
6. DESA: The World's Cities in 2016. UN, p. 27 (2016)
7. Driscoll, J., Teh, B.: The potential of reflective practice to develop individual orthopaedic nurse practitioners and their practice. J. Orthop. Nurs. **5**(2), 95–103 (2001)
8. Gianni, F., Divitini, M.: Technology-enhanced smart city learning: a systematic mapping of the literature. IxD&A **27**, 28–43 (2016)
9. Gianni, F., Mora, S., Divitini, M.: IoT for smart city learning: towards requirements for an authoring tool. In: Proceedings of the First International Workshop on Smart Ecosystems Creation by Visual Design Co-Located with the International Working Conference on Advanced Visual Interfaces (AVI 2016), CEUR-WS, Bari, Italy, vol. 1602, pp. 12–18, June 2016
10. Gil-Garcia, J.R., Pardo, T.A., Nam, T.: Smarter as the new urban agenda: a comprehensive view of the 21st century city. In: Smarter as the New Urban Agenda: A Comprehensive View of the 21st Century City, vol. 11, pp. 1–17. Springer (2015)
11. Gil-Garcia, J.R., Pardo, T.A., Nam, T.: What makes a city smart? identifying core components and proposing an integrative and comprehensive conceptualization. Inf. Polity **20**(1), 61–87 (2015)
12. Habitat, U.: Urbanization and Development Emerging Futures. World Cities Report (2016)
13. Hallnäs, L., Redström, J.: Slow technology-designing for reflection. Pers. Ubiquitous Comput. **5**(3), 201–212 (2001)
14. Hemment, D., Townsend, A.: Smart Citizens, vol. 4. FutureEverything Publications, Manchester (2013)
15. Khansari, N., Mostashari, A., Mansouri, M.: Impacting sustainable behavior and planning in smart city. Int. J. Sustain. Land Use Urban Plann. (IJSLUP) **1**(2) (2014)
16. Klecha, L., Gianni, F.: Designing for sustainable urban mobility behaviour: a systematic review of the literature. In: Citizen, Territory and Technologies: Smart Learning Contexts and Practices, vol. 80, pp. 137–149. Springer (2018)
17. Krogstie, B.R., Prilla, M., Pammer, V.: Understanding and supporting reflective learning processes in the workplace: the CSRL model. In: European Conference on Technology Enhanced Learning, pp. 151–164. Springer (2013)
18. Lam, D., Head, P.: Sustainable urban mobility. In: Energy, Transport, and the Environment, pp. 359–371. Springer (2012)

19. Lidman, K., Renström, S.: How to Design for Sustainable Behaviour? A Review of Design Strategies and an Empirical Study of Four Product Concepts. Chalmers University of Technology (2011)
20. Marsh, J., Molinari, F., Rizzo, F., et al.: Human smart cities: a new vision for redesigning urban community and Citizen's Life. In: Knowledge, Information and Creativity Support Systems: Recent Trends, Advances and Solutions, pp. 269–278. Springer (2016)
21. Mora, S., Gianni, F., Divitini, M.: Tiles: a card-based ideation toolkit for the internet of things. In: Proceedings of the 2017 Conference on Designing Interactive Systems, DIS 2017, pp. 587–598. ACM, Edinburgh (2017)
22. Nam, T., Pardo, T.A.: Conceptualizing smart city with dimensions of technology, people, and institutions. In: Proceedings of the 12th Annual International Digital Government Research Conference: Digital Government Innovation in Challenging Times, pp. 282–291. ACM (2011)
23. Nielsen, J.: Usability 101: Introduction to usability. Nielsen Norman Group (2003)
24. Oates, B.J.: Researching Information Systems and Computing. Sage, London (2005)
25. Petersen, S.A., Canova-Calori, I., Krogstie, B.R., Divitini, M.: Reflective learning at the workplace-the MIRROR design toolbox. In: European Conference on Technology Enhanced Learning, pp. 478–483 (2016)
26. Pettersen, I.N., Boks, C.: The ethics in balancing control and freedom when engineering solutions for sustainable behaviour. Int. J. Sustain. Eng. 1(4), 287–297 (2008)
27. Ratti, C., Townsend, A.: The social nexus. Sci. Am. 305(3), 42–49 (2011)
28. Reiersølmoen, M., Gianni, F., Divitini, M.: DELTA: promoting young people participation in urban planning. In: Conference on Smart Learning Ecosystems and Regional Development, vol. 80, pp. 77–89. Springer (2017)
29. Siegel, M.A., Beck, J.: Slow change interaction design. Interactions 21(1), 28–35 (2014)
30. Wood, W., Quinn, J.M., Kashy, D.A.: Habits in everyday life: thought, emotion, and action. J. Pers. Soc. Psychol. 83(6), 1281 (2002)

Smart Learning – Involving People in Design

People-Centered Development of a Smart Learning Ecosystem of Adaptive Robots

Daniel K. B. Fischer, Jakob Kristiansen, Casper S. Mariager,
Jesper Frendrup, and Matthias Rehm[✉]

Department of Architecture, Design and Media Technology,
Aalborg University, 9000 Aalborg, Denmark
matthias@create.aau.dk

Abstract. Robots are currently moving out of the laboratory and company floor into more human and social contexts including care, rehabilitation and education. While those robots are usually envisioned as a kind of social interaction partner, we suggest a different approach, where robots become adaptive tools for facilitating social interaction and learning in special needs education. The paper presents a people-centered design and development process of such a system that is rooted in the close collaboration between the developers and a network of users and caregivers.

Keywords: Smart learning · Adaptive robots · Brain damage

1 Introduction

It has been argued earlier [7] that the concept of Smart Learning Ecosystems (SLEs) with its focus on more informal, personalized, and experiential learning is especially beneficial for learners, who might be challenged by formal educational settings like we can find them in the traditional classroom.

Based on this argument, we started a collaboration with a neurocenter for adolescents with congenital brain damage. Residents stay for three years at the facility in order to follow an individual rehabilitation and education program. Residents have their own apartments and are surrounded by peers and interdisciplinary teams (teachers, therapists, social and health care personnel).

Based on ethnographic studies at the center, we concluded that smart learning ecosystems in this specific context must tailor to the following objectives: cognitive development (e.g. conveying knowledge or reasoning skills), physical development (e.g. training motor skills like moving an arm), and social skills (e.g. scaffolding social interaction or collaborative tasks). After discussions with management and staff it was decided that the physicality of the interaction and of the system itself would be highly important, leading to initial ideas of including robots for creating a new learning experience.

© Springer International Publishing AG, part of Springer Nature 2019
H. Knoche et al. (Eds.): SLERD 2018, SIST 95, pp. 85–98, 2019.
https://doi.org/10.1007/978-3-319-92022-1_8

Ideally, such a robotic learning environment would relate to all three objectives (cognitive, motor, and social). This paper describes the people-centered development process with staff and residents at the neurocenter that aimed at this outcome.

2 Background

2.1 Cerebral Palsy

Cerebral Palsy (CP) is an umbrella term covering a large spectrum of lifelong movement and posture disorders often accompanied by a range of other disorders. At one end of the spectrum, the movement disorders of CP involve mild spasticity and contractions in a single arm and leg restricted to one side of the body. At the other end of the spectrum all four extremities are involved and the people affected suffers from dyskinesia, severe spasms and scoliosis, making wheelchairs a necessity to obtain mobility. The severity of CP can be rated according to a number of classification systems, including the Gross Motor Function Classification System (GMFCS). All adolescents participating in the development of the ecosystem classify as level IV of the GMFCS - Expanded & Revised (GMFCS - E&R). Adolescents belonging to level IV of the GMFCS - E&R are generally dependent on wheeled mobility. They may be transferred in manual wheelchairs or use powered wheelchairs for independent mobility. However, in indoor settings some are able to walk short distances given physical assistance, or use a body support walker [2].

The quality of life (QoL) of adolescents with CP was analysed by Colver et al. based on self-reported data. In the analysis, QoL of people with CP was related to level of impairment and compared to that of adolescents without disabilities. The impairment severity varied greatly among participants although the majority was mildly to moderately impaired. Nonetheless, participants were scattered across all five classification levels of the GMFCS. Feeding possibilities, communication difficulties, existence of seizures and intellectual impairment also varied.

When compared to adolescents in the general population, the QoL of adolescents with CP was better in five domains: Moods and emotions, self-perception, autonomy, relationships with parents, and school life. However, in the domain of social support and peers[1] it was worse. QoL was worsened with increasing impairment in four domains: Moods and emotions, autonomy, social support and peers, and social acceptance. Impairments on the ability to walk and/or an IQ below 70 were factors associated specifically with a reduction of QoL in the domain of social support and peers. A previous study on QoL of children with CP did not reveal any significant variation in QoL in the domain of social support and peers between children with CP and able-bodied children. For this reason, Colver et al. recommends that an effort should be made to help children maintain and create social relationships as well as participate in society as they go through adolescence [5].

---

[1] Social support and peers - assessed by the social support available from friends and peers.

2.2 Robots in Rehabilitation and Education

Robots are used extensively in rehabilitation contexts. Two main approaches can be distinguished, robots for rehabilitation of motor function and robots for training social interaction skills. The latter kind is mainly researched in relation to autistic children. It has been shown that citizens with autism feel more secure when interacting with computers or robots presumably due to their more predictable behavior [10]. The goal is to create a safe space, where the children can train social interaction skills with the robot and hopefully be able to transfer these skills to interactions with other people.

Robots are also extensively used in education, where we can distinguish between (a) robots for learning how to program (e.g. [12]), (b) robots in the classroom as tutors/teachers (e.g. [11]; [4]), (c) robots as motivators (peers) (e.g. [8]). All studies indicate that the physicality of the interaction with the robot system increases engagement and motivation for the students.

We intend to create a game that makes use of robots. Most research on robots in games focuses on creating robotic game partners to play e.g. rock-paper-scissors [1], or 20 questions [9], using the NAO instead of a human interlocutor. Often the focus is on social aspects of communication like context-dependent affective behavior [3]. Another approach can be called robot assisted play and focuses on making physical play available for disabled children. Kronreif describes one example of a robot helping in stacking Lego bricks [6].

The concept of robot-assisted play comes closest to our approach. Discussions with staff and management at the neurocenter made it clear that the goal is not to substitute the human partner in a game but quite contrary to facilitate social interaction between adolescents, which have challenges in initiating such social interactions themselves. Thus, we are suggesting robots that serve as tools for informal learning and participation for all citizens by utilizing the possibility of adaptive robots, i.e. robots that adjust to the capabilities of the users that control them. In our approach, robots are an adaptive tool to play the game but (seemingly) do not have an active role in the play.

3 People-Centered Development Process

The ecosystem was developed around the capabilities of five adolescents with cerebral palsy. The five adolescents, referred to as participant 1–5, were selected in collaboration with a physiotherapist at the neurocenter due to their relatively similar capabilities. A sixth participant took part in the project's final field test.

Results from workshops with the participants played a major role in the design of the ecosystem, but interviews with personnel and existing literature also influenced the development. A total of four workshops and two interviews, one with a physiotherapist and one with a teacher, were conducted. Development of the system took place in parallel with the workshops which served the purpose of observing and analyzing the capabilities of the participants or introducing and evaluating features of the system. After each workshop the system was changed according to the results of that workshop. Even though it was the goal to test with all participants in every workshop, scheduling did not always allow for it.

3.1   Workshop 1

Participants: Participant 2, 3 and 4 took part in all workshop activities while participant 1 and 5 only were present for parts of the workshop. Multiple members of the personnel assisted in the workshop activities as well.

Goal: This workshop set out to acquire information on the capabilities of the participants' motor functioning which should work as a basis for the development of controllers for the robot cars of the system.

Procedures: Two activities were observed and taken part in, namely physical education (PE) and lunch. PE involved carving of pumpkins.

Results: PE and lunch showed that all were able to perform gross movements with their arms, some with more control than others, but the fine motor control of the hand and fingers was generally lacking in some way. Participant 1 could eat by himself but not without spilling. The hand of participant 2 was generally crumpled and she had trouble holding on to a spoon. Furthermore, her concentration was easily challenged by surrounding disturbances. Participant 3's movements were often involuntary and affected by spasms while participant 4 was able to hold a spoon and eat with a fork but had limited control over individual fingers. Participant 5 was only briefly present during lunch and little was observed, but he showed that he was able to move around independently using a powered wheelchair controlled by a joystick.

Reflections: The results of the workshop showed that fine motor control should not be required of the users of the system as the participants struggled in this area.

3.2   Workshop 2

Participants: Participant 1 through 5 all took part in the workshop which was also supervised by a physiotherapist following the development of the project.

Goal: After initial observation of the participants' abilities in workshop 1, the second workshop was arranged with the main goal of introducing the participants to the first prototype of the robot car and two different controllers. The workshop should also investigate whether the participants enjoyed controlling the robot car.

Procedures: The participants were introduced to the car one at a time and was given the opportunity to try out the two controllers. It was decided to make two controllers to provide a challenging but manageable controller option for all participants despite of varying capabilities. The first controller was based on a big push button that did not necessarily demand fine motor control from the user as it could be actuated using the arm or hand. When controlled by this button, the robot car would move in circles by default and drive forward while the button was pushed. The second controller consisted of a joystick that enabled participants to move the robot car freely while controlling the speed as well. Participants received an introduction to the controllers and the robot car before attempting to control it, and afterwards they were asked about their experience with each controller and the robot car in general.

Results: The area in which the participants were able to control the robot car was not restricted in any way which resulted in many problems with cars colliding with or getting stuck under furniture, and because of this the car also occasionally escaped the visual field of the participants. A small joystick was used for the first prototype due to limitations on available hardware at the time, which clearly was too fragile and ill-suited to the grip of the participants. This added confirmation to results from the previous workshop where it was found that controllers should not rely on fine motor control of the user. Moreover, the responsiveness of the joystick was too great for most participants and participant 1 and 5 especially would push the joystick in unintended directions. Multiple participants got confused about the orientation of the robot car resulting in reversed interpretations of forward/backward and right/left directions.

According to the participants themselves, they all enjoyed controlling the robot car and the controllers' ease of use were rated highly and almost equally by participant 1, 2 and 3 while participant 4 thought of the joystick as being easiest to use and participant 5 rated the button as being easiest to use.

Reflections: A lot of useful design improvements could be deduced from the participants' first exposure to the robot car and its controllers. The area of operation of the car should be restricted, the equipment should generally be more durable and the joystick specifically should be larger. Further changes to the joystick should be made to avoid that participants push it in unintended directions, and the orientation of the car should be conveyed more clearly.

Additionally, it was worth noting that the ratings of the controllers and their ease of use was not always in line with observations made by the group.

3.3 Workshop 3

Participants: Four out of the five participants took part in this workshop alongside the physiotherapist.

Goal: The robot car and its controllers had been updated according to the reflections from last workshop and the main purpose of the workshop was to test the participants' ability to control the car after these updates.

Procedures: One by one, the participants were tasked with driving the robot car to two different cones. This was done twice with each controller, totalling four attempts with each controller for each participant. If the participants did not reach the cone within 1 min and 30 s, the time was not recorded. Instead, they received assistance in reaching the cone before continuing with next attempt. After each participant had finished using a controller he/she was asked about the controllers' ease of use.

Results: Participants were generally happy and excited about the workshop. According to the results, all participants reached the cones faster when using the joystick. In some attempts, the physiotherapist assisted the participant before 1:30 had passed and other attempts were interrupted by other members of the personnel. These attempts were considered invalid. The controllers were rated equally or higher compared to last workshop by all but one participant as participant 2 gave a lower rating of the button controller.

Reflections: The results of the workshop indicated that the updates to the controllers had improved the participants' ability to control the robot car. However, observations of the participants also revealed areas where improvements could still be made. Specifically, the joystick should provide better feedback and time spend on changing from one controller to another should be decreased as time spend on this proved to be very tiresome for some participants.

3.4   Workshop 4

Participants: All of the five participants took part in the fourth workshop. The workshop was supervised by the neurocenter's teacher as the physiotherapist from previous workshops was not able to attend.

Goal: Considering that the system had been modified since last workshop and all participants were present, the main purpose of this workshop was to repeat the test of last workshop. Additionally, the workshop should investigate which controller the participants preferred to use. Most notable modifications from last workshop included a new four-position joystick and a physical playing field to restrict the robot car's operational area.

Procedures: The workshop was similar to workshop 3. Participants took part in the workshop individually and was, similarly to workshop 3, instructed to hit the cones twice with each controller. The button controller was the first to be used followed by the joystick. The controller not in use where kept out of sight to avoid distractions and questions were asked after completion of an activity.

Results: Calculation of average times of valid attempts showed that, contrary to last workshop, participants were faster when using the button. However, when asked, participant 1, 2 and 4 all preferred using the joystick. Participant 5 was the only one who preferred the button. It should be mentioned that even though participant 3 was present for the workshop, she quickly got very frustrated causing her to leave the workshop.

Reflections: It was interesting to see that the preferred controller of the participants was not necessarily the one they performed best with, Fig. 1 shows this. The workshop also showed that participants' ability to adapt to a given speed of the car varied a lot. The ability to control the car is necessary to successfully use the system for which reason it should be considered important that the speed of the car can be adjusted to the abilities of the individuals.

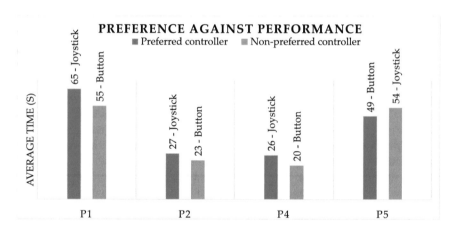

Fig. 1. Shows the average time in seconds used per person for driving into a correct cone while using preferred and non-preferred controllers.

3.5 Interviews

Throughout the project, two interviews with personnel at the neurocenter were conducted; one with a physiotherapist and one with a teacher. The physiotherapist was asked about the citizens' social life where she informed that they need help and incentive to be social which is why the neurocenter promoted social activities such as physical education and theme parties to accommodate social living. In addition, a clear requirement concerned with assignment of personnel for use of the system was expressed, saying that it should not be required to have multiple members of the personnel or particularly technically qualified personnel assigned to use the system. The interview with the teacher contributed with insight into educational activities, content and challenges. It became known that the general educational level at the neurocenter ranged from preschool to second grade and in regard to content the teacher informed that math class often involved simple financial calculations or activities concerned with the clock e.g. by relating time on an analog watch to a digital watch. Furthermore, in Danish class citizens were often tasked with correlating images to text. When the citizens gave answers to questions and exercises it was commonly done by pointing to a picture or word, which was troublesome for some citizens due to spasms. Despite of difficulties answering questions, the citizens generally preferred not to receive help as independence was valued, which the neurocenter also respected.

3.6   System Design Recommendations

The workshops and interviews with personnel provided a basis for the following recommendations to the ecosystem:

- Workshop 2 specifically, made it clear that some participants can be rough when using the controllers. For this reason the controller equipment must be particularly durable and withstand stress from use.
- As the workshops showed that the participants easily lost sense of the robot car's driving orientation, the design of the car should clearly indicate the forward and backward directions.
- In the workshops participants occasionally drove outside the testing area and into obstacles causing the robot car to get stuck. To avoid that the personnel will have to fetch the robot car it should be restricted to a certain area.
- Throughout the workshops it became evident that the optimal speed of the robot car varies between participants for which reason it should be adjustable.
- Interviews with personnel made it clear that the system should be functioning with the presence of no more than one member of the personnel.
- User friendliness is important and all of the personnel should be able to manage the system.
- Conventional educational activities mentioned in the interviews should provide a basis for the educational content of the ecosystem.

4   Robots for Learning

4.1   The Robot Game

The game consisted of remotely controlled robot cars, three cones placed in a confined space, and a game unit controlled by a supervisor. The cones were equipped with a screen and sensors enabling them to tell when hit by the robot car. The robot game can be seen in Figs. 2 and 3.

The game was a question and answer type game where the user is asked a question which can be answered by driving into the cones showcasing different answers. A question would be asked and the participants would have to drive to the cone showing the correct answer. The questions were made with the input from the teacher at the facility and a big emphasis was placed on the citizens wanting to learn the clock. To accommodate this a game where they would

Fig. 2. Shows three cones and two robot cars in the confined space.

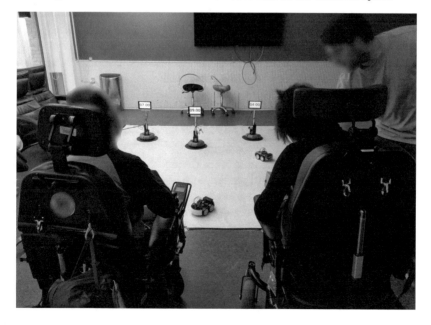

Fig. 3. Shows how the game looks from the users perspective.

receive a picture of an analog watch and then drive to the cone showing the corresponding digital watch were made. Similar games with colors and animals were also made as the teacher told that some citizens needed to learn this. This also increased the variety of the game. In these games the participants would receive a name of either a color or an animal which they had to match to a picture of the corresponding color or animal shown on one of the cones. Additionally the premise of matching analog and digital clocks or text with pictures was told to be a preferred teaching method as mentioned in the interview. Use of the controllers, which was essential for playing the game, could also provide beneficial training in regard to physical rehabilitation.

The game started with a supervisor choosing a game category. Before the category was chosen the cones would display a question mark. After the category had been chosen the game unit sent different answers to the cones. Only one cone would display the correct answer. The supervisor was informed what cone had the correct answer and thereby what question to present to the user.

The users then had to navigate the robot cars to the cones by the use the controllers. If a user hit a wrong cone with the robot car the cone would display a red "X", play a buzzing sound and go back to showing a question mark. If the cone with the correct answer was hit the cone would show a green correct sign, play a "bling" sound and go back to showing a question mark. After the correct cone was hit the game would restart and wait for a new game category.

An example of what the screens on the cones showcased during a game cycle is shown in Fig. 4.

4.2   Field Test

Participants: Participant 1, 4, 5 and 6 attended the field test. All participants had followed the developmental workshops except participant 6 who only had been introduced to the system once. The physiotherapist who had been following the majority of the workshops was present as well.

Goal: The goal of the field test was to introduce a multiplayer aspect of the system and observe how the participants interacted during game play.

Procedures: The four participants were divided into two teams consisting of participant 4 and 5 as team 1, and participant 1 and 6 as team 2. The two participants playing together received a random controller each, either the button or joystick controller, which was used to control each individual car. After having played 3 games the participants would swap controllers. When the participants had tried each of the controllers they where then asked if they would like to continue playing. This would be repeated until the participants said no or other scheduled activities interfered. Because of the possibility of one person being significantly better than the other the game was presented as a cooperative game to avoid discouraging any participants who might be unable to hit the cone before the other participants. When the team was done playing they would be asked individually to rate the game.

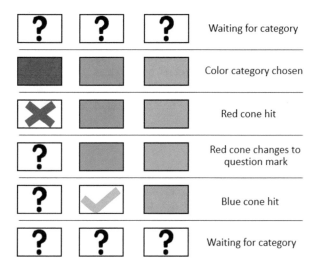

Fig. 4. Shows a game cycle from the color category. The correct answer is blue in this example. The player first hits the incorrect cone showing red and after that hits the correct cone showing blue.

Results: As it can be seen in Fig. 5, Team 1 played 39 times after which the session ended because of other scheduled activities. Even though participant 5 only managed to hit the correct cone first 23% of the time, with an average time of ~15 s, he remained engaged in the game throughout the test and initiated many interactions with participant 4, e.g. by discussing the correct answer. Participant 4 hit the correct cone first 77% of the time, with an average time of ~19 s.

Team 2 played 9 times after which the session ended due to participant 6 struggling to control the car. Participant 1 and 6 had little to no interactions with each other, which possibly could be attributed to the fact that participant 6 needed assistance to control the car as this entailed that most of participant 6's communication was directed to the person helping her. Participant 1 hit the correct cone

		Team 1		Team 2	
		Participant 4	Participant 5	Participant 1	Participant 6
Games	Joystick	19	20	4	5
	Button	20	19	5	4
Hit	Joystick	13 (~68%)	3 (~15%)	3 (~75%)	4 (~80%)
correct	Button	17 (~85%)	6 (~32%)	1 (~20%)	1 (~25%)
cone	Total	30 (~77%)	9 (~23%)	4 (~44%)	5 (~56%)
Avg. success time		~19s	~15s	~108s	~88s
Rating		5	5	4	5

Fig. 5. Shows how many times each participant hit the correct cone first, which controller was used when hitting the cone, the average time used for hitting the correct cone and the participants' rating of the experience.

first 44% of the time with an average time of ∼108 s, where Participant 6 hit the correct cone first 56% of the time, with an average of ∼88 s.

All of the participants except participant 1 rated the experience with a 5 on a scale from 1 to 5, where 1 was bad and 5 was good. Participant 1 gave a rating of 4. Conversations with the physiotherapist after testing also revealed that the system could be particularly helpful in the physical rehabilitation.

5   Discussion

Throughout the development of the project it became evident that while a people-centered development process can contribute with a lot of useful information to take into consideration when developing a system it can also be challenging. As briefly mentioned in Sect. 3 it could be challenging to schedule workshops where all participants were available which hampered testing and data collection. Furthermore, environments for testing would normally have to adapt to the availability and preference of participants which meant that testing was carried out at the neurocenter where control of the environment was limited. Because of this, interrupts from other members of the personnel were experienced several times further hampering testing. The mood of participants should be taken into account when testing. For some participants it could change rapidly which would impact test results and their willingness to participate in workshops. In workshops that set out to quantitatively measure performance of participants when they were controlling the robot car a trade-off often arose between quality of measured data and user enjoyment. Specifically, whenever participants got stuck or lost track of the objective in a particular test it was often decided to assist them, either by us or by the supervising staff member, to maintain their enjoyment of participation. However, at the same time this made collection and comparisons of data difficult as testing conditions seldom were identical.

Workshops as well as the field test was followed by short question sessions to gain opinions on single components or the game as a whole. In the course of obtaining rapport with the participants these question sessions were carried out by the experimenters and it is possible that some participants have given particularly positive ratings to please the experimenters. In comparison of observations and ratings it was also evident that the two did not always line up for which reason the ratings by themselves does not suffice for evaluation of the systems feasibility. However, results from the field test were promising, especially in the aspect of promoting social interactions, and participants seemed to enjoy the game, but further testing has to be conducted in order to conclude on the potential benefits of the system.

6   Conclusion

In this paper we described how a people-centered design and development process resulted in a learning ecosystem, where adaptive robots are tools for facilitating social interaction in informal learning games. Our claim is that this type of innovative system design only becomes possible with the inclusion of the whole context in which the system is envisioned to work. In our case this included apart from the actual users, i.e. the adolescents living at the center, their therapists and teachers as well as management of the center. With the system in its current state, it becomes possible to work closer now with the learning agenda at the institution to transform existing content and develop new content for the new learning ecosystem. We have argued above, that ideally such a robotic learning environment would relate to all three institutional learning objectives (cognitive, motor, and social). With the prototypical games presented during the field test, it is obvious that cognitive learning is easily supported by the setup. For the motor learning, we have positive comments from the physiotherapist that can see increased motivation for training with joystick or button interface. In relation to social learning it remains to be shown in a long-term evaluation which effect the robotic learning ecosystem has on social interaction between the adolescents.

Acknowledgments. We would like to thank all citizens, staff members, and management for their cooperation and dedication in this project.

References

1. Ahn, H.S., Sa, I.K., Lee, D.W., Choi, D.: A playmate robot system for playing the rock-paper-scissors game with humans. Artif. Life Robot. **16**(2), 142 (2011). https://doi.org/10.1007/s10015-011-0895-y
2. CanChild: gross motor function classification system-expanded & revised(gmfcs-e&r), Feburary 2018. https://www.canchild.ca/en/resources/42-gross-motor-function-classification-system-expanded-revised-gmfcs-e-r
3. Castellano, G., Leite, I., Pereira, A., Martinho, C., Paiva, A., McOwan, P.W.: Context-sensitive affect recognition for a robotic game companion. ACM Trans. Interact. Intell. Syt. **4**(2), 1–25 (2014)
4. Chin, K.Y., Hong, Z.W., Chen, Y.L.: Impact of using an educational robot-based learning system on students' motivation. IEEE Trans. Learn. Technol. **7**(4), 333–345 (2014)
5. Colver, A., Rapp, M., Eisemann, N., Ehlinger, V., Thyen, U., Dickinson, H.O., Parkes, J., Parkinson, K., Nystrand, M., Fauconnier, J., Marcelli, M., Michelsen, S.I., Arnaud, C.: Self-reported quality of life of adolescents with cerebral palsy: a cross-sectional and longitudinal analysis. Lancet **385**(9969), 705 – 716 (2015). http://www.sciencedirect.com/science/article/pii/S0140673614612290
6. Kronreif, G.: Robot Systems for Play in Education and Therapy of Disabled Children, pp. 221–234. Springer, Berlin (2009). https://doi.org/10.1007/978-3-642-03737-5_16
7. Krummheuer, A.L., Rehm, M., Lund, M.K.L., Nielsen, K.N., Rodil, K.: Reflecting on co-creating a smart learning ecosystem for adolescents with congenital brain damage. In: Mealha, Ó., Divitini, M., Rehm, M. (eds.) Citizen, Territory and Technologies: Smart Learning Contexts and Practices, pp. 11–18. Springer International Publishing, Cham (2018)
8. Kory, J.M., Jeong, S., Breazeal, C.L.: Robot learning companions for early language development. In: Proceedings of ICMI, pp. 71–72. ACM Press (2013)

9. Pardo, D., Franco, Ò., Sàez-Pons, J., Andrés, A., Angulo, C.: Human - humanoid robot interaction: the 20q game. In: Bravo, J., López-de Ipiña, D., Moya, F. (eds.) Ubiquitous Computing and Ambient Intelligence, pp. 193–199. Springer, Berlin (2012)

10. Robins, B., Dautenhahn, K., Boekhorst, R.T., Billard, A.: Robotic assistants in therapy and education of children with autism: can a small humanoid robot help encourage social interaction skills? Univ. Access Inf. Soc. 4(2), 105–120 (2005). https://doi.org/10.1007/s10209-005-0116-3

11. Saleh, A.R.A., Abdelbaki, N.: Innovative human-robot interaction for a robot tutor in biology game. In: Proceedings of the 18th International Conference on Advanced Robotics (ICAR), pp. 614–619. IEEE Press (2017)

12. Sooraksa, P., Adn Anurak Jansri, S.S., Klomkarn, K.: Tree robot: an innovation for steam education. In: Proceedings of the 2016 IEEE International Conference on Real-time Computing and Robotics, pp. 338–341. IEEE Press (2016)

A System of Innovation to Activate Practices on Open Data: The Open4Citizens Project

Nicola Morelli, Amalia de Götzen[(✉)], and Luca Simeone

Aalborg University, A. C. Meyers Vænge 15, 2450 Copenhagen, Denmark
ago@create.aau.dk

Abstract. The increasing production of data is encouraging government institutions to consider the potential of open data as a public resource and to publish a large number of public datasets. This is configuring a new scenario in which open data are likely to play an important role for democracy and transparency and for new innovation possibilities, in relation to the creation of a new generation of public services based on open data.

In this context, though, it is possible to observe an asymmetry between the supply side of open data and the demand side. While more and more institutions are producing and publishing data, there is no public awareness of the way in which such data can be used, nor is there a diffuse practice to work with those data.

The definition of a practice for a large use of data is the aim of the Open4Citizens project, which promoted initiatives at different levels: at the level of immediate interaction between citizens, experts and open data, at the level of the creation of an ecosystem to work with data and at a level that could support the institutionalisation and consolidation of the new practice.

Keywords: Open data · Innovation · Citizens participation

1 Introduction

The increasing number of government initiatives for the publication of open data is generating an important information resource, which is also incremented by a technological trend that multiplies the number of devices that are recording different aspects of human life, natural environments or urban contexts [1, 2]. Since 2009, when President Obama issued the first executive order to publish all the government information that does not need to be kept secret, a number of government initiatives have started in USA, followed by UK, Kenia, India, Singapore, Mexico, Russia and Europe [3, 4].

The aim of such a large diffusion of initiatives was to increase the government transparency (citizens access to government data), to support service development by third parties (typically the smart city approach) and to develop a new generation of services that stimulate the economy [4, 5].

The increasing relevance of open data as a resource for innovation immediately showed the potential for improving the quality of services offered to citizens in their everyday life: services that could facilitate wayfinding, shopping, transportation or

© Springer International Publishing AG, part of Springer Nature 2019
H. Knoche et al. (Eds.): SLERD 2018, SIST 95, pp. 99–109, 2019.
https://doi.org/10.1007/978-3-319-92022-1_9

healthy habits. Innovation in such aspects could in turn generate a larger innovation system involving local business, public administrations, organisations or interest groups [6].

The present situation, though, is characterised by an asymmetry between the supply side and the demand side. The policies for opening data have been focusing on the implementation of the datasets repositories, rather than on the re-use of them, whereas the long-term demand-side still needs to be adequately stimulated. This is mainly due to a) the lack of a consistent framework to orchestrate and assess strategic interventions to shape an open data ecosystem [4] and b) the absence of a consolidated practice - and a community of practice - that exchanges knowledge and experiences while working with open data [7].

An open data ecosystem includes a range of activities, not only related to the release and publication of open data sets, but also to the treatment and the interpretation of these data, all the way up to the development of pathways showing directions for the usage of open data [6]. Of course, an ecosystem should also be defined by the actors, and the political and organisational infrastructure promoting or participating to those activities.

The activation of such an ecosystem would, in fact, be the basis for a profitable use of open data; however, the activation of such resource would also need a system of innovation [3]. In other words, making data fluid and available is a necessary but not sufficient condition to activate this resource; a learning process needs to be started, which increases the awareness of the opportunities offered by open data. An exploration is needed among citizens, government agencies, private stakeholders and other actors, in order to deeply understand the potential of this resource.

Kapoor et al. [3] observe that in the current paradigm the exploration of the potential and the value realisation is left to civic hackers, developers, small business and entrepreneurs. This is a limitation in the definition of the problem space: these actors are in fact the solution owners, i.e. the people that are most able to generate technical applications using open data, but they often lack an overview of the issues they are trying to address. The inclusion of problem owners - i.e. citizens, public administrators and interest groups that have a clear view of critical problems to solve - would instead call for an open and broader process, based on participation and co-creation.

Given those premises, in order to support the use of open data, Kapoor et al. propose a structured system of innovation, consisting in three subsystems:

- A system of records, including datasets and the treatment needed to make data usable;
- A system of insights, including tools, algorithms and APIs, which would allow for data to be visualised or used in apps and services;
- A system of engagement that would generate social and collaborative capabilities.

The construction of such system would make it possible to support an innovation process that would involve social actors that are usually unfamiliar with open data and unaware of their potential. The creation of a community of practice should consist of a learning-by-doing process, which means that learning a practice of working with data is possible through a real involvement of a community in the creation of solutions at different technology readiness levels, from concepts to operating services. According to Wenger [8], a community of practice is a community of people that engage in a shared process of collective learning within a shared domain. Their involvement is not necessarily intentional, that means that it is unlikely that people will come together to learn how to use open data, but they will possibly join their efforts and spend their energy to solve cogent problems related to their community.

The Open4Citizens project is a good exemplification of this innovation ecosystem. The project is, indeed, generating the elements of the subsystems described by [3] as it includes a data repository (a platform), a system of insights (perspectives and inspiration on how to use data) and a system of engagement (hackathons). This paper will look at Open4Citizens as a system of innovation, particularly focusing on how this system supported learning processes and strengthened the relation of citizens and other actors regarding the access to and use of open data.

2   The Open4Citizens Project

Open4Citizens (04C) is a European project supported by a funding scheme oriented toward Cooperative Awareness Platforms for Sustainability and Social Innovation (CAPSSI).

The project, started in 2016, is generating opportunities for citizens, interest groups, municipalities and local businesses to get better insights, inspiration or support to develop projects based on the use of open data.

The genesis of the project started from the concept of hackathons as co-creation and participatory processes. Traditionally, hackathons are a well-known strategy to accelerate innovation, by grouping IT experts in a "pressure cooker" event, which in few intense days can produce innovative solutions [9].

The presence of IT experts in hackathons is giving a relatively high certainty to develop interesting solutions; the absence of possible users of the hackathons' outcomes, however, also implies an equally high possibility that such solutions do not match real and concrete problems. This motivates the O4C team's idea to extend the participation to their hackathons and involve actors with different knowledge and expertise, in the perspective to engage problem and solution holders in an intense co-creative process. The involvement of citizens and other relevant stakeholders in activities that use open data is also a way to activate a process of learning-by-doing, in which such stakeholders will be able to figure out the full potential of open data by participating in the creation of a new generation of public services.

This strategy, though, changes the whole conception of hackathons and their organisation. Especially in the early days, hackathons were self-organized gatherings, where a group of people (typically, IT experts) would meet for 24 or 48 h and work on issues of common concern. The organization of such early hackathons was quite loose and spontaneous [9]. Conversely, the hackathons of O4C required a long preparation process, which is needed to make sure that an ecosystem of relevant actors is gathered in the hackathon event: this means a long preparatory work before the hackathon and a post-hack process (Fig. 1).

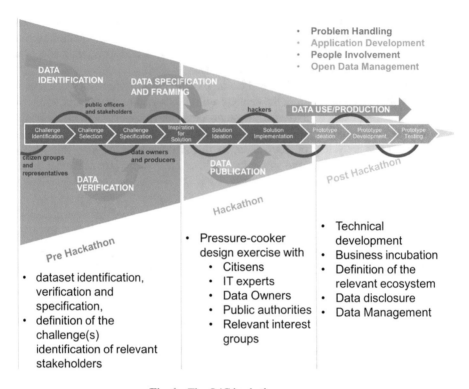

Fig. 1. The O4C hackathon process.

The hackathon event is obviously still the central part of this process: like in the format of the previous hackathons, the O4C events consist of two or three days intense working time, in which participants are collaborating in groups, with periodic presentation of their progress at the end of each day. The final presentation usually includes a review by an invited jury, which often selects a group to be granted with an award. The presence of non-IT-skilled people requires that these events have to be accurately planned and facilitated, also using inspiration tools (e.g. inspiration cards showing possible uses of open data), templates (to map users needs and to outline service journeys), and specific tools and exercises to learn how to deal with data, like data cards (Fig. 2). The O4C team collected these facilitation tools in a preliminary hackathon starter kit and in a Citizens' hackathon toolkit.

Finally, the post-hack phase is the phase in which the hackathon outcomes are tested, incubated and validated. This phase includes an intense process of incubation, development and integration of the hackathon's outcome into the existing administrative, technical and economic systems that constitute public services. This part of the hackathon process implies intersections with political, business related and technical issues, that often obstruct the innovation process, but sometimes accelerate it. The success in this phase very much depends on the presence of key actors in the hackathon ecosystem that have promoted and followed the process.

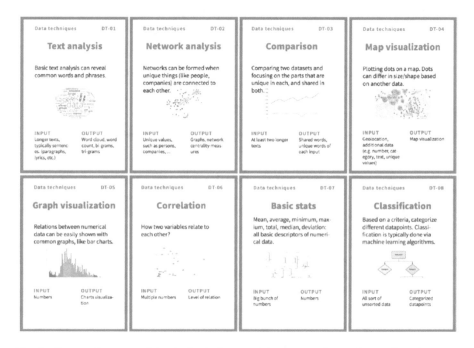

Fig. 2. Data cards are a tool for the hackathon participants to understand open data, figure out techniques to work with them and negotiate a practice in a group of hackathon participants [10].

3   The Outcome of the Open4Citizens Project

The work on the hackathon events and on the whole hackathon cycle highlights the need to organise the innovation process on open data around an ecosystem that collects the relevant stakeholders in relation to a process and create compressed and accelerated innovation sessions.

At the same time, it is important to make sure that the knowledge acquired in the organisation of such innovation sessions is consolidated in an infrastructure that would make the hackathon experience reproducible and facilitate the public access to data. The project team defined this infrastructure as an OpenDataLab, i.e. a virtual platform and possibly a physical reference point for open data use and diffusion. The research effort of O4C is therefore concentrated on different perspectives, as in a matrioshka (Fig. 3):

- the hackathon event, i.e. a "pressure cooker" innovation process, of the duration of 2–3 days;
- the hackathon process, i.e. a process of 6–7 months that gathers the fundamental components of an open data ecosystem in respect to a specific problem area;
- An OpenDataLab, i.e. a permanent innovation place, where citizens can learn about open data, even outside the hackathon process.

Fig. 3. The "matrioshka" model defines three logical levels in the process of activating the use of open data [10]

Those three levels also represent a progressive extension of the community of practice [8] around open data from IT experts to a broader community of citizens, public administrators, interest groups and small business. The hackathon is one of the most common practice to generate fast innovation processes within the community of IT experts and coders. Another way of looking at this matrioshka model for O4C is as a progressive learning process where the various actors learn how to use open data and where the practice of using open data is activated.

Opening the hackathon event to a broader social context requires a process that defines an ecosystem of actors, tools and new practices. The hackathon process is aimed at defining a balanced interaction of three elements:

- People: the process has to make sure that the relevant people are present in the hackathon, which may play a role in the development of a solution. Among them, it is important to involve the relevant issue owners, such as public administrators that are looking for innovative public services, or even data owners, which are often unaware of the potential of the datasets they own or are eager to give value to those data as a resource for innovation.
- Data: the hackathon process should make sure that relevant data are collected before the hackathon, in order to offer the participants the raw material to work on during the hackathon event.
- Challenges: the process of creating a new practice is only possible if the participants are genuinely involved in the solution of concrete problems. This means that relevant challenges have to be proposed, for which the participants will be willing to spend their energy and time. The choice of the challenge often refers to urgent political issues, social emergencies or organisational issues.

Finally, the OpenDataLabs represent a consolidation and institutionalisation of the new practice. In the O4C view, they are public innovation places [11] that support the dissemination of a culture of open data, offer services (e.g. consultancies, facilitation for hackathons, working tools) for those who want to use open data, and actively promote any initiative for the use of this resource. In other terms, they are learning spaces where a variety of actors can gather and explore the potential and challenges of open data. The initial inspiration for the OpenDataLab comes from maker spaces: besides being public spaces where people are able to manipulate material to generate innovative solutions, maker spaces are also places to exchange knowledge and disseminate a practice of digital fabrication. Likewise, OpenDataLabs are supposed to play the same social and institutional role when manipulating open data.

In consideration to Kapoor's elements for an innovation system [3], the effort of the O4C team consisted in identifying such elements at the three levels mentioned above. This defines a systemic strategy for supporting the generation of new practices at all levels (Table 1). It is worth noticing that the extension of the focus from the technical treatment of open data to the social context requires a broader interpretation of Kapoor's terms, to include social and practical issues related to the participation to the hackathon event, the hackathon process and the activities in the OpenDataLab.

Table 1. Elements of the innovation ecosystems at the three levels of the O4C project.

	System of records	System of insights	System of engagement
Hackathon event	Datasets, data repositories	Visualisation and inspiration tools	Hackathon and workshop facilitation, data sprints
			Involving data owners and issue owners
Hackathon process	Shared issues, community culture and attitudes, institutional settings	Activities focused on data, people and challenges	Hackathon event, participatory activities
OpenDataLab	Knowledge about open data	Calls for projects, call for data, fund raising, navigation and visualisation support	Activities that engage citizens, public administrators, data owners and other relevant stakeholders

3.1   The Hackathon Event Level

The system of records at the hackathon level consists in the collection of all the raw material for the hackathon activity. The raw material for a creative activity on open data is, of course, a number of datasets: links to the most relevant data repositories were collected for the participants before the hackathon event.

The system of insights in the hackathon event consists of the visualisation and inspiration tools provided to the participants. This includes inspiration cards, examples and visualisations of existing datasets.

The system of engagement consists of the various tools and strategies for engaging participants, including facilitation or data sprints. It is worth noticing that the engagement of participants also depends on non-technical issues, such as the participation of data owners or issue owners (e.g. public administrations proposing a problem to solve or awards for the best project).

3.2   The Hackathon Process Level

The raw material in the hackathon process consists of the information coming from institutional, social and organisational frameworks. It includes shared issues, community culture and attitudes, institutional settings, including laws, hierarchical structures and regulations. In other words, the raw material for the hackathon process is the social, technical and organisational ecosystem in which certain problem areas can be addressed with the use of open data.

The system of insights in the hackathon process consists of the activities that are supporting the formation of the hackathon ecosystem, they include activities focused on the three main components of the hackathons: challenges (What are the relevant problems that the hackathon should solve?), data (Which datasets can be relevant? Which ones are available? Which ones can be found from different sources?) and people (Who are the people that would be motivated to solve the challenges?)

Finally, the system of engagement consists of the participatory activities during the hackathon process, including the hackathon event and a number of other preparatory or post-hack events, including data sprints, service jams, meetings and workshops.

3.3   The OpenDataLab

The raw material collected in the OpenDataLabs is the whole knowledge around open data, that means: an archive of datasets, applications, inspiration and networking tools that would make it possible for any user of the Lab to use data, work with data or get information on how to use open data.

The OpenDataLabs are also supposed to collect and support activities on open data, such as calls for projects, calls for data and fundraising opportunities, data navigation and visualisation support. All these elements are part of the system of insights related to the OpenDataLab.

The number of activities that engage citizens, public administrators, data owners and other relevant stakeholders in open-data-related processes represents instead the system of engagement in the OpenDataLab.

4   Discussion

The creation of a practice of designing with open data is a process of social construction that has an important learning dimension and that aggregates an ecosystem of people, technologies, data, institutions, and challenges. The hackathon is certainly an effective tool to support such a process, especially if it is based on the construction of the ecosystem in the pre-hack phase and followed by a solid support to the development of the outcomes in the post-hack phase. This temporal articulation (pre-hack, hackathon and post-hack phases) goes beyond the typical duration of such events (24–48 h [9]). The hackathon per se is an accelerated and compressed learning process that works very well to raise the awareness of the potential of open data, but is often not sufficient to consolidate the process of learning that the generation of a practice would need. A logical tension emerges, between the quick and intense process in the hackathon and what takes for normal citizens to get acquainted with data tools and methods.

This makes the extended temporal articulation of the 04C hackathon processes and, in particular, the OpenDataLab more relevant for the creation of a practice of working with data. The matrioshka model for O4C is a progressive learning process that culminates with the OpenDataLab as a more stable learning space where various actors can tinker with data, data visualisation processes and data handling tools and can experiment with related facilitation and organisational capabilities. In other words, the OpenDataLab would be fundamental for the institutionalisation of a practice around open data. In this sense, OpenDataLabs can be seen as public innovation places [11] for the consolidation of such practice.

The three levels mentioned in this paper are, therefore, configuring a complex structure that invests different aspects of the construction of a practice. The hackathon event is the moment in which interaction and co-creation happen. It is the level of effective participation of multiple stakeholders, such as citizens, public authorities, interest groups, data owners and business companies. It is sometimes a highly emotional event, because of the intense and concentrated work it requires. The event is generating awareness, opening perspectives, introducing to new tools, creating short circuits among actors that would otherwise never have a chance to meet or work together. All the participants to the hackathon event have an opportunity to get closer to a practice of working with open data, although the event per se is too short to create consolidated knowledge.

The hackathon process is dedicated to the construction of an ecosystem. It is a moment of intense negotiation between different stakeholders, organisations and institutions. This is a process of definition of the challenges and identification of relevant datasets. What the stakeholders of this process learns in this period is to recognise the relevant players related to specific challenges and specific datasets. The learning process concerns systemic aspects of working with open data.

Finally, the OpenDataLab represents the consolidation of knowledge, the creation of a shared pool of skills, capabilities and opportunities in a community, the creation of knowledge that can influence the institutional framework, public policies and governance of open data.

The 04C team intends to propose OpenDataLabs as a seed for a movement of democratisation of open data, getting inspiration from the FabLab movement. In the last decade, FabLabs, maker spaces and personal fabrication labs have created a new culture and practice of production, which is opening new perspectives for material production and is promising to democratise the access to production resources, including the creation and support of commons [12, 13]. This movement, as well as the OpenDataLabs, in the intention of the O4C team, will create new institutional frameworks for the production of material goods and services for citizens. Within this perspective, open data can be considered as new commons [14, 15].

5   Conclusion

This paper used the construct of system of innovation to look at the Open4Citizens project and, in particular, focused on how this system supported learning processes and strengthened the relation of citizens and other actors with regard to the access to and use of open data.

The potential of open data to become a resource for society has been quite clearly perceived by governments and organisations. However, there is a lack of distributed awareness of how to use this resource. Citizens, public authorities and institutions still do not know what to do with open data and how to work with them. While the supply side becomes more and more relevant with the creation and distribution of new datasets in many areas, the demand side is still underdeveloped. While the practice of analysing and exploring data is maturing, thus creating systems of insights [3] in the private and public sector [16], the last step in the learning process that would activate open data as a commons is still far from being done: there is no culture of designing with data, and no practice of dealing with it and, therefore, the system of engagement [3] that would activate open data as a resource is still to be developed.

The Open4Citizens project illustrated in this paper is proposing a strategy to close this gap. The system of innovation illustrated in this paper is an attempt to implement a structured learning process (pre-hack, hackathons, post-hack with OpenDataLabs) to build a community and to activate a practice that does not yet exist.

References

1. Shepard, M. (ed.): Sentient City: Ubiquitous Computing, Architecture, and the Future of Urban Space. MIT Press, Cambridge Mass (2011)
2. Ciuccarelli, P., Lupi, G., Simeone, L.: Visualizing the Data City. Springer, Milan, Heidelberg, New York, Dordrecht, London (2014)
3. Kapoor, S., Mojsilović, A., Strattner, J.N., Varshney, K.R.: From open data ecosystems to systems of innovation: a journey to realize the promise of open data. Paper presented at the Bloomberg Data for Good Exchange Conference, New York City, 28 September 2015
4. Turki, S., Martin, S., Renault, S.: How open data ecosystems are stimulated? In: Paper presented at the Proceedings of the International Conference on Electronic Governance and Open Society: Challenges in Eurasia, St. Petersburg, Russian Federation (2017)
5. Janssen, M., Charalabidis, Y., Zuiderwijk, A.: Benefits, adoption barriers and myths of open data and open government. Inf. Syst. Manag. **29**(4), 258–268 (2012)
6. Huijboom, N., Van den Broek, T.: Open data: an international comparison of strategies. Eur. J. Pract. **12**(1), 4–16 (2011)
7. Zuiderwijk, A., Janssen, M., Davis, C.: Smart government, citizen participation and open data. Inf. Polity **19**(12), 17–33 (2014)
8. Wenger, E.: Communities of Practice: Learning, Meaning, and Identity. Cambridge University Press, Cambridge, New York (1998)
9. Seravalli, A., Simeone, L.: Performing hackathons as a way of positioning boundary organizations. J. Org. Change Manag. **29**(3), 326–343 (2016)
10. Kun, P.: Design case study report. Open4Citizens Deliverable 2.3 (2017). http://open4citizens.eu/

11. Manzini, E., Staszowski, E.: Public and collaborative: exploring the intersection of design, social innovation and public policy (2013). http://www.designagainstcrime.com/files/publications/pub_2013_public_and_collaborative.pdf
12. Bollier, D.: Think Like a Commoner: A Short Introduction to the Life of the Commons. New Society Publishers, Gabriola Island (2014)
13. Hess, C.: Mapping the new commons. In: Paper Read at International Association for the Study of the Commons. University of Gloucestershire, Cheltenham (2008). https://papers.ssrn.com/sol3/papers.cfm?abstract_id=1356835
14. Morelli, N., Mulder, I., Concilio, G., Pedersen, J., Jaskiewicz, T., Götzen, A.D., Aguilar, M.: Open data as a new commons: empowering citizens to make meaningful use of a new resource. In: Paper Presented at the 4th International Conference on Internet Science, Thessaloniki (2017)
15. Morelli, N.: Open data: creating communities and practices for a new common. In: DSI Workshop on Digital Technologies to Support Social Innovation, Thessaloniki, Springer (2018)
16. Davies, T.: Open data, democracy and public sector reform. M.sc dissertation, University of Oxford (2010)

Urban Innovation Through Co-design Scenarios

Lessons from Palermo

Enza Lissandrello[1]([⊠]) [iD], Nicola Morelli[2], Domenico Schillaci[3], and Salvatore Di Dio[3]

[1] Department of Planning, Aalborg University, Rendsburggade 14, 9000 Aalborg, Denmark
enza@plan.aau.dk
[2] Department of Architecture, Design and Media Technology, Aalborg University, Rendsburggade 14, 9000 Aalborg, Denmark
nmor@create.aau.dk
[3] PUSH Design Lab, Piazza Sant'Anna 3, 90133 Palermo, Italy
{d.schillaci,s.didio}@wepush.org

Abstract. This paper aims to contribute to current research on learning through designing for urban innovation. It provides a framework methodology for a multidisciplinary ecosystem as a participatory method developed in the context of Mobility Urban Values (MUV), an EU Horizon 2020 project (2017–2020), that addresses the issue of behavioral change towards (more) sustainable mobility lifestyle in EU cities. The paper frames the MUV method through the combination of theories on collaborative urban planning and participatory design with a background rooted on governance of public participation, as the interplay between co-creation (thick participation) and co-design (thin participation). MUV participatory method is envisioned as a learning infrastructure that engages at different levels communities, citizens, and stakeholders. This paper addresses the question on how enabling urban innovation through sensitive phases of sociological and technical components to produce learning. The conceptual background of the MUV method and the first application phase of co-creation/co-design for the old city center neighborhood in Palermo, Italy, provide lessons on the results of this approach for a future research agenda and the loop learning. While the method is adopted specifically in relation to mobility urban values, MUV method can inspire a variety of other cases questioning urban innovation through socio-technical learning.

Keywords: Participatory methods · Capacity building · Loop learning

1 Introduction

Participatory methods for urban innovation constitute resources for building social capital and system-thinking learning, through the exchange among diverse perspectives, knowledge, and experiences in local contexts. Learning, through participatory methods, implies a systemic perspective in which the confrontation with initial assumptions of participants, the creation of meanings, the mindset of people involved, sometimes open up opportunities for capacity building and making urban places [1]. This paper explores

© Springer International Publishing AG, part of Springer Nature 2019
H. Knoche et al. (Eds.): SLERD 2018, SIST 95, pp. 110–122, 2019.
https://doi.org/10.1007/978-3-319-92022-1_10

the linkage between learning-by-design and urban innovation. It draws from a conceptual background that combines studies on urban governance and participatory design [2–8] to advance a new urban participatory method that includes different degrees of thick and thin participation [9] into an ecosystem of loop learning [10–12]. This method has been envisioned within the context of Mobility Urban Values (MUV), a Horizon 2020 EU research and innovation project, aimed at shaping an innovative urban policy development to improve livability and health conditions in EU cities. This paper discusses the main concepts and approaches that have emerged in MUV research on participatory methods for urban innovation with an empirical exploration of the first co-creation/co-design phase in Palermo, Italy (December 2017).

In this context, urban innovation is understood in relation not just to the 'invention' of a solution-design, but as a generative capacity building [4] a learning capacity that develops through the travelling of ideas among designers, planners, citizens as individuals and communities, local businesses and stakeholders and policy administrators. Urban innovation entails thus a systemic 'translation from the level of conscious actor invention and mobilization to that of routinization as accepted practices, and beyond that to broadly accepted cultural norms and values' [4].

The background of urban innovation so intended is a participatory process with a focus on "the activities by which people's concerns, needs, interests, and values" can possibly be incorporated into "decisions and actions of public matters and issues" [9] (p. 14). In discussions on participatory methods, a great part is on the purpose. In urban studies, Innes and Booher [2] famously frame five purposes of participation as (1) getting knowledge about the public interest, (2) improving decisions by using the local knowledge of citizens, (3) promoting and achieving fairness and justice, (4) securing legitimacy for decisions, (5) meeting the legal requirements of the process.

The two categories of thick participation (involving groups of citizens) and thin participation (involving citizens as individuals) illustrated by Nabachi and Leighninger are at the base of the MUV participative ecosystem that develops through a double level of co-creation as community building [5, 13] and co-design as infrastructuring [7, 8]. MUV addresses a new participatory method that involves together thick and thin levels of participation to enhance systemic learning within the current European urban democratic structures of deliberation [3]. Scholars have shown that bottom-up initiatives have the chance to succeed allowing direct and active involvement of citizens, by design, in community development [14]. However, the issue of effective policy through citizen participation cannot stand from bottom-up initiatives alone. It depends from actions of facilitation of a systemic process that involves together in a co-creation and co-design diverse actors at diverse levels. The facilitation of systemic processes is therefore a key for urban planners and designers: 'who facilitate public participation, as well as amongst whom, when and to what degree' [14] (p. 266). Facilitation is understood as a form and function of the MUV ecosystem participatory method; MUV as a 'learning infrastructure' iterates and aims to develop loop learning [12] among participants, planners and designers. Loop learning is represented and 'materialized' through scenarios.

The remaining of this paper is structured in three main parts. The following part illustrates the theoretical framework of the MUV method as based on urban studies and participatory design studies. It follows the case of Palermo and a conclusion that discusses the findings in terms of loop learning and potential principles that emerge from MUV method for urban innovation. While MUV participatory methodology is, at the moment of writing, explored, implemented and tested in several EU cities (Helsinki, Ghent, Fundao, Amsterdam, Barcelona, and Palermo), the paper builds on the first experience of MUV method in Palermo historical city center.

2 MUV Participatory Method: An Urban Ecosystem of Co-creation and Co-design

Three main concepts are at the base of MUV participatory method and research: capacity building, infrastructuring and loop learning. Capacity building and infrastructuring develop from collaborative urban planning and participatory design approaches [13, 15] here interpreted respectively within two diverse levels of thick participation (involving groups of citizens) and thin participation (involving citizens as individuals) [9]. These two levels intertwine and overlap together to generate loop learning [10–12, 16]. In MUV, these loops are represented by scenarios specifically oriented to the issue of mobility urban values (mobility scenarios, service scenarios, and scalability scenarios (Fig. 1).

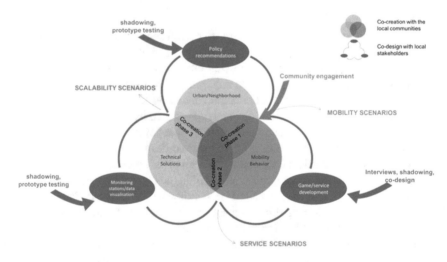

Fig. 1. Representation of the MUV participatory ecosystem with the interplay of the two levels of co-creation and co-design aligned on three loop-learnings with outcomes on mobility scenarios, service scenarios and scalability scenarios

2.1 Co-creation: Capacity Building and Thick Participation

Capacity building as thick participation is at the base of the conceptual understanding of 'co-creation' in an urban context; here understood as a process-oriented to long-term effects developed by interaction among citizens and communities. Capacity building, in the MUV context, consists in the facilitation of a set of moments and structures of opportunities that bring the resources of groups of citizens and individual citizens into new ideas, strategies and actions of urban concerns [1] This process of capacity building relates to three dimensions (1) the knowledge resources, (2) the relational resources, (3) the capacity of mobilization [17]. In MUV, these dimensions define co-creation as within resources and capacity of urban neighborhood. In this way, the community engagement and facilitation of resources sharing as well as new capacities occur by the consideration of challenges, needs and values collected with and through citizens and communities' imaginaries of future. A focus on possibilities-building can have effects on network creation and resource sharing that can last well beyond the temporality of design solutions. These effects become important in view of developing urban policy through scalability strategies and long-term processes for urban innovation and governance.

2.2 Co-design: Infrastructuring and Thin Participation

Infrastructuring as thin participation is the conceptual grounding of 'co-design' as a dynamic in which local communities, here understood as individual citizens, engage in dialogic moments and interaction, imagining and constructing a set of structured visions [15]. Infrastructuring is here understood as a process that facilitates the emergence of new design opportunities as resulted, adapted and re-arranged in relation to specific circumstances. This process can allow stakeholders not just to find a solution but to generate trust and to develop new design approaches that can connect individuals, small organizations and established institutions [6, 8, 18]. In MUV, infrastructuring is interpreted as distributed over three different degrees of co-design as; (1) design with individual citizens as a diffuse design experience related to creativity props, i.e. scenarios, prototypes, and games; this supports the citizens' capability to generate personal value (utility) by aggregating material (i.e. services, technologies, tools) and immaterial resources (i.e. knowledge, social links, word of mouth); (2) co-design with formal institutions supporting value creation processes that can develop on specific design of services, platforms, software applications, and games that might transform the knowledge-interaction between users and providers; (3) co-design with aggregates of social ecosystems, i.e. a set of relevant stakeholders that identifies elements that strategically can be extracted, communicated and scaled-up in relation to the possible impact of the service function and use.

In MUV participatory ecosystem (Fig. 1) the interplay between co-creation (capacity building) as thick participation (involving groups of citizens) and co-design (infrastructuring) as thin participation (involving citizens as individuals) generate loop learning as potential orienting scenarios (mobility scenarios, service scenarios and scalability scenarios). Figure 1 represents the two levels of co-creation and co-design within the MUV ecosystem.

The level of co-creation develops through three diverse phases by considering the diverse themes related to MUV: urban neighbourhood, mobility behaviour, and technical solutions. The co-creation level consists of three phases (1) the knowledge resources of urban neighborhood by consideration of mobility behavior challenges, needs, and values (co-creation phase 1); (2) the relational resources enhanced by community engagement and by pooling diverse kinds of communities in the consideration of common challenges in mobility within the current constraints of technical solutions (co-creation phase 2) and (3) the capacity of mobilization of new ideas for technical solutions, strategies and actions of urban concerns through open imaginaries of future (co-creation phase 3).

The level of co-design develops in concert with the three phases of co-creation: (1) the first focuses on the service design development derived by the citizens' capacity to generate personal value (utility) in relation to possible design orienting mobility scenarios dimensioned and refined through processes of interviewing and shadowing; (2) the second focuses on the co-design with formal institutions supporting value creation processes on specific service orienting scenarios by knowledge-sharing and interaction between users and providers and facilitated through shadowing, prototyping, and testing; (3) the third shapes co-design with a set of relevant stakeholders through monitoring and possible data visualization and through shadowing, prototyping and testing for the identification of elements that strategically can be extracted, communicated and scaled-up for future policy recommendation and urban innovation. The outcomes of the interacting phases of co-creation and co-design aim to constitute specific loop learning scenarios.

2.3   Scenarios as Representation of Loop Leaning

The final aim of the MUV participative ecosystem is a loop learning approach to urban innovation and urban policies. The three phases that align co-creation and co-design represent three loop-learning phases as inspired by systemic levels of learning (adapted from organizational learning literature [10–12]. The three scenarios - mobility scenarios, service scenarios and scalability scenarios - represent the diverse learning processes enabled by the MUV participatory ecosystem as a 'learning infrastructure' [12] (p. 295). Inspired by single-loop learning [10], this occurs 'whenever an error is detected and corrected without questioning or altering the underlying values of the system' while double-loop learning occurs 'when mismatches are corrected by first examining and altering the governing variables and then the actions' [12]. There is also another type of learning, defined as 'deutero-learning' which 'can occur by going *meta* on single or double-loop learning' [16] (p. 1179 our Italic). Three questions are interpreted to trigger diverse loop learning scenarios whenever phases of co-creation and co-design interplay, intertwine and overlap at time within MUV participatory method. MUV scenarios represent learning as in the form of questions: 'are we doing things right (single-loop learning)? Are we doing the right things (double-loop learning)? Can we participate in making well-informed choices regarding strategy, objectives, etc. (e.g. triple loop learning)?' [11] (p. 452). This set of questions represent an inspiration to scenarios and not a simple and direct adaptation from theory to practice.

Mobility Scenarios as Single Loop Learning

Mobility scenarios are understood in MUV participatory ecosystem as a result of a process of 'single loop learning' in which co-creation and co-design dynamics based on thick and thin participation align on the question 'are we doing things right (in relation to our desires of urban mobility in the neighborhood)?' This question simplifies a simple learning process with effect on adaptation of community mobility behavior e.g. in relation to individual travel behavior. Mobility scenarios are inspired by a loop-learning scenario of possible individual corrective actions with a possible impact on the neighborhood livability.

Service Scenarios as Double Loop Learning
Service scenarios are understood in MUV participatory ecosystem as a result of a process of 'double loop learning' in which co-creation and co-design dynamics based on thick and thin participation align on the question: 'are we doing the right things (in relation to possible services of urban mobility within the current constraints of existing technology)? This question is as an inspiration for a collective participatory reframing of the 'problem' of mobility from individual to systemic, including mobility technical solutions to be exploited, prototyped, and tested in urban contexts with effects as shaping values.

Scalability Scenarios as Triple Loop Learning
Scalability scenarios are understood in MUV participatory ecosystem as a result of a process of 'meta-learning' (e.g. triple loop learning). This means that co-creation and co-design dynamics – as based on thick and thin participation – can align their focus on the question: 'Can we participate in making well-informed choices regarding strategy, objectives, etc. in urban mobility? In the MUV context, this question is seen as an inspiration for participants (citizens, local communities, public authorities) for developing new processes of re-framing mobility urban solutions into a set of strategic elements which involve values-creation with possible impacts on existing urban policy (e.g. on urban/mobility/management and digitalization).

3   MUV Participatory Method in Palermo: The Interplay Between Co-creation and Co-design

Palermo[1] is the fifth most populated metropolitan area in Italy, with about 1,2 million inhabitants. The historical center is one of the pilot areas of the MUV project; it extends for almost 250 hectares with high-density. It is the location of the majority of the touristic sites of the city. The area faced several modifications during the last century as seriously damaged during II World War, with a consequential heavy shrinking in population between the 1960s to 1970. Important initiatives for urban generation culminated just in 1990s with the approval of a detailed Executive Plan for historic center. New urban investments mobilized real estate commercialization of historical buildings and cultural entrepreneurship. With the restoring of architectural heritage and the opening of several bars, pubs, restaurants, and touristic attractions and cultural centers, today the historical center is catalyzing again the public attention. In 2013–2014 Palermo municipality has adopted a general urban traffic plan that has integrated previous urban mobility policy. Especially relevant for the urban revitalization of the historical center has been the

---

[1] Palermo is historically an important city, with a population around 850 000 and a vaster surrounding metropolitan area in the Sicilia region.

pedestrialization policy and the recent institutionalization of the Traffic Limited Zone (ZTL) in 2016.

In 2012–2015 a project called Traffic02 won the call "Smart Cities and Communities and Social Innovation" promoted by the Ministry of Education, University and Research in Italy. The project tackled the problem of Palermo's urban traffic, transport, and mobility from another point of view than usual solution-oriented projects for new infrastructures, public transport or hard policies.

One of the scopes of the Traffic02 project was to shift the debate on changing cities by switching from the urban structure (the hardware) to the changes that can be induced by citizens through a change of their behavior and the urban communities' habits (the software). TrafficO2 was thought as an info mobility decision supporting system that tried to foster a modal split through gaming policies and tangible incentives for each individual citizen when making (more) sustainable mobility choice [19] with the help of smartphone applications. Local businesses jointed through an ICT platform (as sponsors) became the stations of a new kind of transport system in which only moving by foot, by bicycle, by local public transport and by car-pooling were rewarded. Each trip from station to station gave O2 points to the user, as the virtual money users can collect to get prizes from the sponsors. A first test of the mobile application (an alpha version) has started during December 2013 with about 80 students selected through a workshop and has shown about the 55% of reduction of CO_2 emissions [20].

Traffic02 experience has constituted a background for MUV research concerning the potential of ICT-based services to enhance urban innovation. MUV participation method includes also local businesses and public authorities' new possibilities of interaction through open data and technology at use as web portals and dashboard that constitute not just tools but new organizational models that tend to flat the relations between citizens and public institutions '…expected to facilitate the construction of an open dialogue with the community and easy access to data already collected and processed and therefore directly "digestible" for citizens and other institutions' [21]. The establishment of these networks can stimulate formal government to provide more efficient mobility services shaped by city-users' expectations.

MUV participative methodology in Palermo is thus shaped by the ambition to enlarge the perspective of the design experience of Traffic02 from a project-design to a process-design based on co-creation (community building) and co-design (infrastructuring). In the case of Palermo, gaming policies and tangible incentives have emerged as successfully elements from the Traffic02 experience. MUV participative ecosystem aims providing a deeper-learning experience for urban innovation through the very process of co-creation (thick participation) and co-design (thin participation) rather than to depart from a project-based approach.

MUV role in Palermo is to facilitate three main processes as: co-creation (as thick participation) and co-design (as thin participation), and solution-oriented loop learning scenarios. Issues as 'sustainable mobility values' and 'technology in use' are not given determinations but shaped through MUV participatory method as a learning infrastructure for urban innovation (Fig. 1).

3.1 MUV Co-creation in Palermo: Capacity Building and Thick Participation

In Palermo, a two-day MUV workshop has engaged groups of citizens and individual citizens into a process of capacity building in which new ideas, strategies, and actions for urban innovation, departing from mobility issues, have been facilitated. Three existing typologies of communities have been respectively and progressively involved as resources of knowledge, relational and mobilization capacity by the MUV team: (1) community of place, as all those group of citizens who live and work in the neighborhood through the existent 'street councils'[2] (2) community of practice, as all those groups of citizens already involved in initiatives related to sustainable mobility, neighborhood's entrepreneurship, cultural or social innovation activities through 'Mobilita Palermo'[3] and Palermo Innovation Network MeetUp[4] (3) community of interest, as those of groups potentially interested in municipal initiatives included private entrepreneurial commercial groups[5].

Fig. 2. Co-creation location Palermo, Garibaldi Theatre

The diverse themes of the MUV participatory ecosystem (urban/neighbourhood, mobility behavior, technical solutions) as described in Fig. 1 have been tackled, following a first phase of co-creation with a workshop in which communities of citizens have been encouraged to reflect on their habitual and desired mobility behaviors in Palermo. This in relation to urban and neighborhood identities, current challenges, needs, values and imaginaries of

---

[2] A community of about 50 people, most of with a deep knowledge of the neighborhood and local business owners (shops, bar, restaurants, bed and breakfasts) usually facilitated by municipal administrators.

[3] http://palermo.mobilita.org.

[4] https://www.facebook.com/groups/inetworkpalermo/. These sites are multi blogs top web-page rank visited in Palermo by citizens.

[5] As uGame in which Palermitan urban planners and game designers join forces facing social issues through gamification approaches.

future. The first co-creation workshop took place at the Garibaldi Theatre, the old symbol of a new Palermo, today in a process of change (Fig. 2).

The MUV community engagement activities in Palermo have been performed both online and offline. The first MUV co-creation workshop has been related to an Eventbrite, a well-known ICT platform in Palermo, to easily share the information about the session and manage participants' venue through a landing page about MUV. The links were shared on Facebook and through direct emails sent to local communities. An offline campaign took place across local locations in the communities mentioned above. Since the first co-creation workshop venue was an open place, MUV posters were hanged at the Garibaldi Theater.

This was another "offline way" of reaching potential participants informing them about the upcoming workshop. The project webpage went online on the 7th of December 2018, as well as the Eventbrite. In one week it registered 93 sessions, 195-page visualization, and 62 users. Through Eventbrite, 307 people were reached and the 30 tickets available went sold out. A mix of consolidated and well-known tool-kits for encouraging communities as the "World Café[6]" or "The Arrow[7]" have been adopted to start-up a dialogue on current people's mobility behavior and desired imaginaries of urban and neighborhood mobility. These tools have enhanced people's reflection as a community of citizens, of adaptive actions to meet their desired imaginary (single learning loops). As a first result, the participants of the co-creation workshop visualized their urban mobility scenarios on how to correct their own mobility behavior towards safer and sustainable one with a positive transformative effect for the neighborhood livability.

Therefore, these groups of citizens provided a collection of possible corrective actions, either immediately achievable or already implemented, which are instrumental to the attainment of desired mobility scenarios. Data collected in this session have been used to select concrete people's mobility experiences in Palermo. The result has been clustered in two different mobility experiences, one called 'the rational' and the other named 'the emotional'. Participants were then requested to list pros and cons of four selected sustainable travel modes (i.e., walking, biking, public transport, carpooling), describing their current modal splits.

A second co-creation workshop activated through a 'mobility brainstorming' aimed to identify groups of citizens' personal positive experience related to a specific travel mode in Palermo. The result was a mind map of the happy/unhappy mobility experience of mobility in the urban environment. The groups of citizens in Palermo, have envisioned two ways to improve mobility within the historical center: an extrinsic one, linked to policies, infrastructures and environmental aspects; and an intrinsic one, related to socio/cultural

---

[6] The 'world cafe' is a flexible design tool to enhance group dialogues; in Palermo, such a dialogue facilitated by the authors on the question on Palermo citizens' challenges and desires in relation to their personal mobility habits and behavior.

[7] The 'arrow' is a design tool to encourage participants to define, decide and eventually achieve their goals starting from their own vision of future. In other words, this tool encourages participants to an imagination based on back-casting, so to imagine a vision of the future and defining the steps of actions that they need to undertake them as an orientation. In Palermo, the arrow has encouraged communities of citizens to shape a common vision of sustainable mobility.

aspects, personal behaviors and motivations. The intrinsic way has been perceived as a faster-track to achieve the desired mobility scenarios.

This co-creation workshop enabled participants to reflect on how their intrinsic motivations play a role making their mobility experiences as positive. Three main variables possibly alter their actions towards the achievement of desired scenarios through behavior change: these variables consist in (1) social motivations as feeling part of a community; (2) intrinsic personal benefits as having an enjoyable experience and (3) monetary rewarding as receiving a real or a virtual recognition in relation to a more sustainable behavior. The higher ranked 'related motivation' in mobility positive experience in Palermo is the social motivation. Mobility is a vehicle for community building (familiarizing with strangers on public transport, organizing with other people to overcome risks as strikes or accidents, joining active groups of walkers or bikers or having contacts with shopkeepers) (Table 1).

Table 1. Positive mobility experiences and related groups of citizens' motivations from the co-creation session in Palermo

Positive experience	Related motivation
Familiarizing with strangers on public transport	Social motivation
Organizing with other people in case of a strike	Social motivation
Joining events/groups of cyclists met by chance	Social motivation
Get to know the shopkeepers and get in touch with them	Social motivation
Sharing long car trips with friends	Social motivation
Carrying shopping bags easily by bike	Intrinsic benefit
Uncovering the beauty of the city and its hidden places	Intrinsic benefit
Being quicker and more punctual thanks to the bike	Intrinsic benefit
Discovering new places and shops simply by walking	Intrinsic benefit
The first cycling route on the new bike lane	Intrinsic benefit
Improving back pain, improving posture	Intrinsic benefit
Sharing gasoline through carpooling, saving money	Monetary reward

Personal motivation resulted as a driver for mobility behavioral change to meet community's desired values as safety and sustainability. In the first co-creation session, technology at use in citizens' everyday life - both within the urban space and through apps - has emerged as an important element in relation to social interaction. In Palermo, nowadays mobility is a source of frustrations for many people one the conclusions and leading design principles coming from co-creation sessions is the need of a playful experiences to have 'fun' again with mobility and the value of social interaction as priority area. Results in terms of capacity building have been shown in relation to mobilization capacity through the connectivity of diverse communities of citizens – sharing the same imaginary and values for the future – of a safe and sustainable urban future.

3.2   MUV Co-design in Palermo: Infrastructuring and Thin Participation

The concept of co-design as thin participation has been operationalized in Palermo through user-centered design sessions involving both mobility management officers and local business owners. Mobility management office sessions have been as 'a day in the mobility management office of Palermo' to collect technical insights on data sources, touchpoints, and decision-making mobility planning practice. Local business-owners have been engaged to reflect on digital equipment, business networks and connections at the neighborhood and urban scales. The specific challenge of the first co-design session was to find solutions to reduce vehicular traffic in the old town especially on nightlife. This challenge includes conflicts among local residents' (families) in need of parking spaces and of a quite family lifestyle that clash with young people (students and professionals) lifestyle and mobility behavior who enliven the neighborhood at night. The co-design with individual citizens has been focused on game prototypes to enhance individual capacities to generate personal design in relation to their game attitudes (i.e. competitive vs cooperative, multi-players vs single players) and in relation to soft or hard materials (i.e. apps; touchpoints, urban spaces). This prototyping has pointed to the social dimension of mobility as a vehicle for improving social interaction and safety of meeting places (walking together, collective transport, car-sharing). Co-design with formal institutions have led to design principles that might transform the knowledge-interaction between users and providers (i.e. through mobility data exchange). MUV co-design as infrastructuring includes also local businesses and public authorities' new possibilities of interaction through gamification; open data and technology at use as web portals and dashboard emerged as potential triggers for new organizational models for new trustful relations between citizens and public institutions, especially important in Palermo.

4   Loop Learning Dynamics

As discussed in the first part of this paper, loop learning has been an inspiring concept in the context of MUV for developing a pathway towards urban innovation processes. Re-framings mobility urban values into a set of strategic elements with a possible impact on existing urban policy (e.g. on urban/mobility/management and digitalization) is one of the expected results of the MUV project. The conclusions of this first phase of co-creation/co-design in Palermo has provided lessons in terms of systemic learning. Taking again the question 'can we participate in making well-informed choices regarding strategy, objectives, etc. in urban mobility? [11] in Palermo, a mobility scenario has been related to personal motivations in co-creation sessions. The starting preconditions in the historical center of Palermo in terms of urban mobility is a network of automobiles roads within urban historical cultural values in the neighborhood and missing significant and effective participatory methods in current urban decision making. The case of Palermo offers interesting principles for urban innovation; the role of MUV as a learning infrastructure is those to generate new relational and mobilization capacity through interactions between existing communities of citizens but also to establish a new channel among citizens, stakeholders, and public administration. Trust is an especially important issue in view e.g. of securing a more transparent use of data on mobility, more safety through social interaction and to meet the legal requirements for a new participative urban culture in decision-making processes in Palermo. Learning from the first phase of MUV co-creation and co-design in Palermo, a new mobility scenario shows the need to turn the negative frustrations of citizens and communities towards their potential responsibility on governing their own mobility behavior and consciously their impact in their urban environment in terms of livability. Playful dynamics among stakeholders, citizens, and administrators have resulted to have an important role whenever a conflict of values occur e.g. between family life and nightlife in the neighborhood, with the hope to turn individuals and communities towards a common desired future of mobility urban values.

Acknowledgments. This research has received funding from the European Union's Horizon 2020 research and innovation programme, under grant agreement No 723521. Our thanks to three reviewers SLERD 2018 for their helpful comments on a first draft of this paper.

References

1. Coaffee, J., Healey, P.: My voice: my place: tracking transformations in urban governance. Urban Stud. **40**(10), 1979–1999 (2003)
2. Innes, J.E., Booher, D.E.: Reframing public participation: strategies for the 21st century. Plann. Theor. Pract. **5**(4), 419–436 (2004)
3. Fung, A.: Varieties of participation in complex governance. Public Adm. Rev. **66**(s1), 66–75 (2006)
4. Healey, P.: Transforming governance: challenges of institutional adaptation and a new politics of space. Eur. Plann. Stud. **14**(3), 299–320 (2006)
5. Cars, G., Healey, P., Madanipour, A., De Magalhaes, C.: Urban Governance, Institutional Capacity and Social Milieux. Routledge, London (2017)
6. Manzini, E.: Design research for sustainable social innovation. In: Design Research Now, pp. 233–245 (2007)
7. Björgvinsson, E., Ehn, P., Hillgren, P.-A.: Participatory design and democratizing innovation (2010)
8. Manzini, E., Jégou, F., Penin, L.: Creative communities for sustainable lifestyles (2008)
9. Nabatchi, T., Leighninger, M.: Public Participation for 21st Century Democracy. Wiley, Hoboken (2015)
10. Argyris, C., Schon, D.A.: Theory in Practice: Increasing Professional Effectiveness. Jossey-Bass, San Francisco (1974)
11. Romme, G.A., Van Witteloostuijn, A.: Circular organizing and triple loop learning. J. Organ. Change Manag. **12**(5), 439–454 (1999)
12. Tosey, P., Visser, M., Saunders, M.N.: The origins and conceptualizations of 'triple-loop' learning: a critical review. Manag. Learn. **43**(3), 291–307 (2012)
13. Healey, P.: The pragmatic tradition in planning thought. J. Plan. Educ. Res. **28**(3), 277–292 (2009)
14. Blanchet-Cohen, N.: Igniting citizen participation in creating healthy built environments: the role of community organizations. Commun. Dev. J. **50**(2), 264–279 (2014)
15. Manzini, E.: Design. When Everybody Designs. MIT Press, Cambridge (2015)
16. Argyris, C.: A life full of learning. Organ. Stud. **24**(7), 1178–1192 (2003)

17. González, S., Healey, P.: A sociological institutionalist approach to the study of innovation in governance capacity. Urban Stud. **42**(11), 2055–2069 (2005)
18. Murray, R., Caulier-Grice, J., Mulgan, G.: The Open Book of Social Innovation. National Endowment for Science, Technology and the Art, London (2010). Young Foundation NESTA
19. Vinci, I., Di Dio, S.: Reshaping the urban environment through mobility projects and practices: lessons from the case of Palermo (2016)
20. Di Dio, S., Rizzo, G., Vinci, I.: How to track behaviours' change towards more sustainable habits: the serious game of Traffic02 (2015)
21. Molinari, A., Maltese, V., Vaccari, L., Almi, A., Basssi, E.: Big data and open data for a smart city (2014)

Smart Learning Resources

Informing Informal Caregivers About Dementia Through an Experience-Based Virtual Reality Game

Jette Møller Jensen, Michelle Hageman, Patrick Bang Løyche Lausen,
Anders Kalsgaard Møller, and Markus Löchtefeld[✉]

Department of Architecture, Design and Media Technology,
Aalborg University, Aalborg, Denmark
{jjen14,mhage14,plause14}@student.aau.dk, {akmo,mloc}@create.aau.dk

Abstract. In 2017 it was believed that nearly 50mio people suffered from dementia. Besides the actual patients, the group that is mostly affected by this disease are informal caregivers. Informal caregivers – people without a formal education in the field of health care – can suffer from severe physical- and mental health issues due to the changing behaviour of the person with dementia. It has been shown that these can be overcome by giving the caregivers information and guidance about dementia at an early stage of the condition. In this paper we present our investigation of an interactive experience-based Virtual Reality game and how it can inform informal caregivers about symptoms of dementia. Our initial exploration demonstrates the potential that such a game holds in supporting informal caregivers.

Keywords: Dementia · Alzheimer's Disease · Virtual Reality
Games · Experience-based learning

1 Introduction

Dementia is an umbrella term describing a condition in which the mental capabilities of a person decreases. One of the most common reasons for this is Alzheimer's Disease. However, there are over 200 different diseases which can cause dementia [13]. Symptoms of dementia can vary, but it often affects cognitive functions such as memory and personality of the patient and increases with age. In 2017 it was believed that nearly 50mio people suffered from dementia and it is expected that this number will rise to 75mio in 2030 [1].

At this point in time there are no cures for the illnesses which causes dementia [13]. Despite this, there are still ways to ease the symptoms. Treatment of dementia can vary, depending on which illness caused it in the first place. It is in any case of utmost importance that the caregivers try to understand the condition and support the person suffering from it [14]. Taking care of a person with dementia (PwD) can have adverse effects on the caregivers [8]. An especially

© Springer International Publishing AG, part of Springer Nature 2019
H. Knoche et al. (Eds.): SLERD 2018, SIST 95, pp. 125–132, 2019.
https://doi.org/10.1007/978-3-319-92022-1_11

important role is taken by the informal caregivers, who are people without formal education in the field of healthcare that take care of a PwD. Informal caregivers are often very close relatives such as spouses or children that live together with the PwD. Unfortunately, taking care of the PwD can cause psychological and physical distress and even lead to depression for the informal caregivers [8]. This is due to the changing behaviour of the PwD and the amount of pressure that the care puts upon the caregivers [5]. However, these issues can be reduced by giving the caregivers more information and guidance about dementia at an early stage of the condition [7]. The World Health Organization therefore stresses the importance of developing new solutions to support informal caregivers [13]. Several prior approaches have successfully explored how experiencing the symptoms of dementia can educate and sensitize caregivers, however they normally require experts to conduct and explain the symptoms [6, 10] or just present a video that provides very little interactivity [2, 15].

To overcome these limitations, in this paper we present our investigation of an interactive experience-based Virtual Reality (VR) game, i.e. a game based on real-life experiences, and how it can inform informal caregivers about symptoms of PwD. The content of the game is based on experiences of informal caregivers living together with PwD, that have been extrapolated from a local support workshop for informal caregivers. While the main focus are informal caregivers, the content of the game is also well suited to educate the general public about dementia.

2   Related Work

While there is often a large variety of information on symptoms and problems of PwD available through national and local organizations, Andrews et al., discovered that the informal caregivers felt they did not have enough knowledge about dementia and that the information was not readily available to them [3]. Therefore, there is a potential that the existing information is not well enough adapted or disseminated for the informal caregivers. Egan and Pot argue that innovative technologies hold a great potential for the education of caregivers about the symptoms of dementia [9]. Especially, courses offered via the internet can be a promising method to inform caregivers [4]. Several trials showed that these interventions can have a positive effect specifically on lowering the stress and increasing the well-being of informal caregivers. Particularly, programs, that combine general information with specific care-giving strategies, were more likely to improve the informal caregiver's well-being [4].

However, understanding the symptoms and the struggles that PwD are going through is a complex problem. One way to improve the understanding of PwD that has been proven to be successful in the past is to experience what it is like to have dementia by taking the perspective of the PwD [10]. With the *Into D'mentia* cabin, Hattink et al. developed a learning experience split into three parts. In this program, the user gets an introduction to the experience, a simulation in a kitchen- or a dinner setting (where the visitor gets to experience it like the PwD), along with a debriefing in which the user discusses his experiences with an expert. In the last part, the user gets to share their experience with other users, in a group situation [10]. They found that this way of helping the informal caregivers has been greatly successful, but as it is using a specially designed cabin, it is only available at specific places which hinders the access for informal caregivers.

A similar experience has been designed by Bevelle called *12 min in Alzheimer's Dementia*. Through several sensory alteration tools which simulate the effects of living with dementia [6] the users would experience the disease first-hand. The users are given special glasses to obstruct their sight, latex gloves with tape, along with a substance in their shoes to impair their mobility functions, and a headset playing indistinct chatter to obstruct their hearing [6]. While very successful in communicating the symptoms, it has the disadvantage that it requires an instructor that moderates the experience.

To overcome the mobility limitations and the need for an instructor, 360° videos have been utilized in the past that would allow the user to take the role of the PwD. *Alzheimer's Research UK* released an application that tries to convey the impact of dementia through the use of short scenarios that are presented in 360° videos in VR [2]. The application does not have a specific target group and gathered a substantial amount of downloads already, however to the best of our knowledge no formal investigation on its effectiveness has been conducted. Wijma et al. took a very similar approach [15]. They designed three 360° simulation videos that depict different situations and levels of interaction with informal caregivers of PwD. These scenarios cover different contents and information related to dementia symptoms e.g. communication problems between the informal caregiver and the user (who takes the role of the PwD). The videos would be shown using a Head Mounted Display (HMD) to informal caregivers. Their evaluation highlighted that after the experience users felt to have deeper understanding of the struggles of PwD. Both approaches did not provide interaction possibilities for the users, this might have decreased the immersion and might have reduced the level of empathy the participants felt with the PwD.

In our work we decided to use VR as well, as it holds a great potential to resemble the experience as closely as possible with a high level of immersion. Furthermore VR has been proven to be effective for learning e.g. in K-12 education [12]. Hayhurst identified several design challenges with respect to designing VR for PwD but not for informing informal caregivers, still he concludes that "VR as training aid to caregivers should help more people understand the needs of PwD which may lead to improved levels of care" [11]. In this paper we want to build on these works and extend them with more interactivity. The user that is trying to relive the experience of a PwD should actively take the role and interact with the same limitations that PwD would face. Making the information more accessible and interactive, would enable to engage the informal caregivers to a higher degree and improve their understanding of living with the dementia.

3   Simulating Dementia Symptoms in Virtual Reality

To create this experience-based VR game we collaborated with a local dementia health-care expert, that regularly holds courses for relatives and informal caregivers. Besides participating in one of these courses we consulted her extensively for the design of the scenarios which the user will experience in the game. As part of the course, different stories of experiences from PwD were disclosed to which the relatives could relate and learn from. While a wide variety of topics and stories were discussed we opted to select two scenarios that would portray two manifestations of dementia symptoms that are often confusing for informal caregivers.

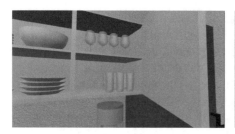 

Fig. 1. The cup scenario of the game; The cupboard is open, but there is no cup yet (left). The NPC has come to the kitchen, to get the cup, which has now appeared (right).

3.1   'Cup' Scenario

In the 'Cup' Scenario, the player will be asked to go get a cup for the Non Player Character (NPC). The NPC represents a relative of the PwD, here their daughter as an informal caregiver. The player will then have to locate the cup in the presented living space. However, the player will not be able to locate the cup. This is done in order to simulate how people suffering from dementia can have problems recognizing everyday objects which are hidden in cupboards or drawers. In the end of the scenario, the NPC will find the cup for the player and act frustrated towards the player. This is done in order to simulate how it can be difficult for relatives to cope with how the person who they once knew are disappearing. Frames from this scenario can be seen in Fig. 1. On purpose the players will in no way be successful, which makes it very frustrating for the players. This was done to represent how the PwD's often will feel very frustrated with themselves due to the disease.

3.2   'Distractions' Scenario

In the 'Distractions' Scenario, the player will be standing in front of the NPC who will start talking to the player. As she talks, there will be other stimuli in the room (dimming lights, increasing volume of the television) to confuse the player. This is done in order to simulate the way that PwD can have problems with handling too much stimuli at the same time. Frames from the scenario can be seen in Fig. 2. In the end of the scenario, the NPC will again become frustrated with the player. As with the 'Cup' Scenario, this is done in order to simulate the ways that relatives can react negatively, since the situation is emotionally difficult for them to handle and understand.

Fig. 2. The *"Distractions"* scenario of the game. The lights has not dimmed yet, and the focus has started to be directed towards the television (left). The lights are dimmed in the scenario and the volume of the TV is increased (right).

After both scenarios a window would pop up which would elaborate on the problems of PwD that led to the scenario. This is in line with the debriefing that is e.g. used in [6,10]. The scenarios demonstrate to the users what it feels like to suffer from dementia and be mistreated by informal caregivers for their cognitive limitations. Along with the scenarios, a waiting area was developed for the game. From this the player could choose one of the scenarios by opening one of two doors. Furthermore, in the area, posters were set up with more information regarding dementia, which the players could read (see Fig. 3).

Fig. 3. The waiting area of the game. The posters can be seen on the walls, and the doors from which the player can enter a scenario can be seen to the right.

The game was implemented using Unity, the 3D models were created using Autodesk Maya. As a VR HMD we used the HTC Vive. And for the VR aspect of the game SteamVR and Virtual Reality Toolkit were used. Furthermore, a tutorial level was included in the game, to demonstrate the basic movement and interaction with objects. The player would move through the environment by teleporting, i.e. they would press a button on the controller, which showed a line to a spot. When they released the button, they would be transported to that spot.

4   Explorative Evaluation

To understand whether this game can be effective in communicating the struggles of a PwD, we conducted an explorative evaluation. While the target group for the game are informal caregivers, it proved to be difficult to get hold of a larger group that could come to our VR Lab. But as the game has the potential to also educate the general population about dementia – which in certain situations e.g. meeting a PwD on the streets that needs help, could also turn into informal caregivers – we opted to test it on students. We recruited one informal caregiver (aged 59, male) and 29 (3 female) students (aged between 20 and 27 years) of our university from which 11 stated that they know or have known a PwD.

For the evaluation the participants would first be greeted and introduced to the game and the procedure. Afterwards, they were asked to fill out a consent form, demographic questionnaire and the pre-study questionnaire. After a short VR introduction and the tutorial level, they found themselves in the waiting room from where they could go through the doors leading to two scenarios. The participants would then play through both scenarios of the game, before answering the post study questionnaire, followed by a semi-structured interview. Overall the procedure would take on average 20 min.

4.1   Results

As part of the pre- and post questionnaire we asked the participants three questions that would require them to self-assess their knowledge about dementia, the challenges, and needs of PwDs on a 5-point Likert scale. The results of these questions indicate that the participants felt to have gained a better understanding of dementia as well as the challenges and needs of PwD (compare Fig. 4). A Wilcoxon signed-rank test for each of the questions revealed a significant difference ($p < 0.05$). Furthermore, we asked participants to list the numbers of symptoms of dementia they knew, before and after playing the game. Before playing the game the participants that were students on average could mention 1.14 symptoms ($S = 0.7$) and afterwards on average 2.48 ($SD = 0.99$).

4.2   Discussion

Overall the results seem very promising that the game was effective in communicating different symptoms of dementia as well as challenges and needs of PwD. However, there is a potential that the increase in knowledge could mainly stem from the explicit information displayed in the posters and at the end of the

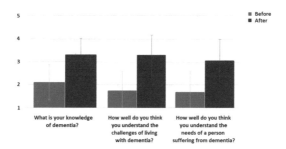

Fig. 4. Results of the user's self-assessment about their knowledge and understanding of PwD before and after having experienced the game.

scenarios on screen. As we did not do a screen recording of the interactions of the participants it is not obvious at this point. However, participants mentioned in the interview that adding more scenarios might improve the game in terms of getting more knowledge about dementia. They also mentioned that the information from the posters should be more incorporated in the scenarios, which lead us to believe that the posters were very helpful in obtaining more knowledge about dementia. The majority of the participants particularly mentioned in the interview that they liked the game even though it was very frustrating but helping in understanding the struggle of PwD. Especially the aggressive NPC was mentioned by nearly all participants, which is in line with the design idea behind it. We wanted the participant to experience how a PwD could feel when living with an informal caregiver that does not understand potential symptoms of dementia.

5   Conclusion and Future Work

In this paper we presented the design and evaluation of an experience-based learning game, that is targeted towards informal caregivers of PwD. Based on two scenarios we allow users to experience what life as a PwD might entail and how negative reactions towards the mental impairments feel like, in order to appeal to their empathy. While our initial evaluation demonstrates the potential that such a game has on how informal caregivers can effectively learn about symptoms of PwD, we are eager to extend on this work. We want to include more symptoms of PwD's and furthermore extend the realism by e.g. also including physical impairments as in [6]. Furthermore we would like to use the SteamVR store and Google Cardboard as ways to make this experience as easily accessible as possible.

References

1. Alzheimer's Diesease International: Dementia statistics. https://www.alz.co.uk/research/statistics. Accessed 3 Feb 2018
2. Alzheimer's Research UK: Virtual reality app offers unique glimpse into life with dementia. http://www.alzheimersresearchuk.org/a-walk-through-dementia-news/. Accessed 3 Feb 2018
3. Andrews, S., McInerney, F., Toye, C., Parkinson, C.A., Robinson, A.: Knowledge of dementia: do family members understand dementia as a terminal condition? Dementia **16**(5), 556–575 (2017). https://doi.org/10.1177/1471301215605630
4. Boots, L., Vugt, M., Knippenberg, R., Kempen, G., Verhey, F.: A systematic review of internet-based supportive interventions for caregivers of patients with dementia. Int. J. Geriatr. Psychiatry **29**(4), 331–344 (2014)
5. Borsje, P., Hems, M.A., Lucassen, P.L., Bor, H., Koopmans, R.T., Pot, A.M.: Psychological distress in informal caregivers of patients with dementia in primary care: course and determinants. Fam. Pract. **33**(4), 374–381 (2016)
6. Clarke, A.: What does alzheimers disease feel like? https://www.theseniorlist.com/2013/07/patented-tool-provides-a-glimpse-into-the-world-of-alzheimers-disease/. Accessed 3 Feb 2018
7. Cooper, C., Balamurali, T., Selwood, A., Livingston, G.: A systematic review of intervention studies about anxiety in caregivers of people with dementia. Int. J. Geriatr. Psychiatry **22**(3), 181–188 (2007)
8. Cuijpers, P.: Depressive disorders in caregivers of dementia patients: a systematic review. Aging Ment. Health **9**(4), 325–330 (2005)
9. Egan, K.J., Pot, A.M.: Encouraging innovation for assistive health technologies in dementia: barriers, enablers and next steps to be taken. J. Am. Med. Dir. Assoc. **17**(4), 357–363 (2016)
10. Hattink, B., Meiland, F., Campman, C., Rietsema, J., Sitskoorn, M., Dröes, R.: Experiencing dementia: evaluation of into d'mentia. Tijdschr. Gerontol. Geriatr. **46**(5), 262–281 (2015)
11. Hayhurst, J.: How augmented reality and virtual reality is being used to support people living with dementia—Design challenges and future directions, pp. 295–305. Springer International Publishing, Cham (2018). https://doi.org/10.1007/978-3-319-64027-3_20
12. Merchant, Z., Goetz, E.T., Cifuentes, L., Keeney-Kennicutt, W., Davis, T.J.: Effectiveness of virtual reality-based instruction on students' learning outcomes in k-12 and higher education: a meta-analysis. Comput. Educ. **70**, 29–40 (2014)
13. Prince, M., Prina, M., Guerchet, M.: Alzheimer's disease international (2013) world Alzheimer report 2013. Journey of caring: an analysis of long-term care for dementia. Alzheimer's Disease International (ADI), London (2013)
14. Van der Roest, H.G., Meiland, F.J., Comijs, H.C., Derksen, E., Jansen, A.P., van Hout, H.P., Jonker, C., Dröes, R.M.: What do community-dwelling people with dementia need? a survey of those who are known to care and welfare services. Int. Psychogeriatr. **21**(5), 949–965 (2009)
15. Wijma, E.M., Veerbeek, M.A., Prins, M., Pot, A.M., Willemse, B.M.: A virtual reality intervention to improve the understanding and empathy for people with dementia in informal caregivers: results of a pilot study. Aging Ment. Health **10**, 1–9 (2017)

ReadME – Generating Personalized Feedback for Essay Writing Using the *ReaderBench* Framework

Robert-Mihai Botarleanu[1], Mihai Dascalu[1,2,3(✉)],
Maria-Dorinela Sirbu[1], Scott A. Crossley[4],
and Stefan Trausan-Matu[1,2,3]

[1] University Politehnica of Bucharest,
313 Splaiul Independentei, 060042 Bucharest, Romania
`robert.botarleanu@stud.acs.pub.ro`, {`mihai.dascalu,`
`stefan.trausan`}`@cs.pub.ro`, `maria.sirbu@cti.pub.ro`
[2] Academy of Romanian Scientists,
54 Splaiul Independenței, 050094 Bucharest, Romania
[3] Cognos Business Consulting S.R.L.,
32 Bd. Regina Maria, Bucharest, Romania
[4] Department of Applied Linguistics/ESL, Georgia State University,
Atlanta, GA 30303, USA
`scrossley@gsu.edu`

Abstract. Writing quality is an important component in defining students' capabilities. However, providing comprehensive feedback to students about their writing is a cumbersome and time-consuming task that can dramatically impact the learning outcomes and learners' performance. The aim of this paper is to introduce a fully automated method of generating essay feedback in order to help improve learners' writing proficiency. Using the TASA (Touchstone Applied Science Associates, Inc.) corpus and the textual complexity indices reported by the *ReaderBench* framework, more than 740 indices were reduced to five components using a Principal Component Analysis (PCA). These components may represent some of the basic linguistic constructs of writing. Feedback on student writing for these five components is generated using an extensible rule engine system, easily modifiable through a configuration file, which analyzes the input text and detects potential feedback at various levels of granularity: sentence, paragraph or document levels. Our prototype consists of a user-friendly web interface to easily visualize feedback based on a combination of text color highlighting and suggestions of improvement.

Keywords: Automated writing evaluation · Textual complexity
Feedback generation and visualization · Natural language processing

1 Introduction

Writing quality is an important component of understanding student success at almost all educational levels. Moreover, providing individual feedback to students on their writings is an essential component for their development; however, it is a resource intensive task and many teachers don't possess the necessary time. Therefore, in most cases, teacher feedback is generally limited during the writing process in most traditional educational scenarios. Thus, computer-based writing systems have been developed to support students' writing development by providing personalized feedback in near real time. These writing systems have the capacity to offer immediate feedback and to keep students more engaged in the writing process.

The aim of this paper is to introduce the beta version of an Automated Writing Evaluation (AWE) system – *ReadME* –, a novel smart learning environment for generating personalized feedback for student writing. Using statistical methods, a wide range of textual complexity indices reported in a large corpus of English writing was reduced to a few principal components which describe quantifiable writing features within that corpus. By comparing component or individual textual complexity scores with the baseline generated from the training dataset, the system is able to generate student feedback that can guide students in improving their writing.

2   Related Work

2.1   Automated Writing Evaluation

According to Roscoe et al. [1], there are two categories of textual evaluation systems that can provide feedback to students: Automated Essay Scoring (AES) systems and Automated Writing Evaluation (AWE) systems. AES systems [2] provide an overall, summative score on an essay and their performance can be evaluated in terms of the correlation between human and automated scores. Second, AWE systems [3] are built on top of AES systems with the aim to provide constructive feedback that goes beyond holistic scoring (i.e., formative in addition to summative scoring). AWE systems are more useful for student development since the received feedback can help them in terms of awareness of potential issues, as well as methods of improvement. Both AES and AWE systems rely on textual complexity indices which provide indications regarding a text's quality or traits of writing style, and can measure different aspects ranging from readability, surface-level numerical metrics (such as paragraph counts, syllable counts, word entropy etc.), to text coherence and cohesion [4, 5].

Many AWE systems have been developed for the writing classroom. Some of them are freely available (in most cases, those developed by researchers), while systems developed by companies usually require a small to medium fee. As an example, *MyAccess!* (https://www.myaccess.com/myaccess/do/log) is a web-based writing system which helps students improve the quality of their essays. The system provides different versions for both students and teachers. With regards to students, *MyAccess!* offers essay templates and access to multiple tools including grammatical advices, spell checker, as well as an automated scoring system. For teachers, the system offers the possibility to monitor students, to create templates, and essay assignments.

WriteToLearn (https://www.writetolearn.net) is another AWE system which helps students enhance their comprehension by teaching them summarization techniques and their writing quality by providing automated feedback. Other AWE systems, which provide automated feedback for students are: *Project Essay Grade* (PEG; http://www. pegwriting.com) or *LightSide Revision Assistant* (https://www.revisionassistant.com/).

2.2   Automated Readability Evaluation

A few systems (e.g., **VisRa** [6] and **T.E.R.A.** [7]) use similar methods to the ones that are meaningfully integrated into our system – *ReadME*, to assess the readability of texts. **VisRa** [6], for instance, provides detailed visual feedback to users regarding the readability of their texts. The addressed facets of readability include linguistic and content-wise appropriateness. During the analysis of a text, the VisRA system uses various indices for measuring readability. These include features such as word length, vocabulary difficulty, nominal forms, sentence length and sentence structure complexity (computed through the depth of a sentences' parse tree). In order to provide an overarching perspective of the provided feedback, the system uses a gradient-based color scheme: blue represents an easy to read text (with dark blue being the easiest to read), whereas hard to read fragments are colored in red, with darker shades meaning that the sentence or paragraph is harder to read. Regarding the provided granularity levels, three different views are integrated: corpus view, block view and a detailed view.

T.E.R.A. [7] is a free tool that computes a text's readability to help students self-assess their products. T.E.R.A. relies on complex features, such as text cohesion. T.E.R.A. can help identify parts of a text which are the most difficult to understand and shows the user which sections need to be improved and why. This can be used by teachers to help students recognize and overcome issues they may encounter in understanding a specific text. The tool provides scores for each of the five components used in the analysis (i.e., narrativity, syntactic simplicity, word concreteness, referential cohesion and deep cohesion) relative to the baseline corpus. Feedback is provided for all of these five components. In contrast to VisRA, the feedback provided by T.E.R.A. is textual and does not indicate issues at sentence or paragraph level, but rather of the entire text. The feedback is more oriented towards teachers by enabling them to analyze whether a text is appropriate or not for a certain class of students.

3   Method

Our goal is to introduce a fully automated method of generating essay feedback in order to help improve learners' writing proficiency that is similar to both VisRa and T.E.R.A. The development of this system (ReadME) is detailed below.

3.1   Selected Corpora

For our initial attempt at developing ReadME, we used the TASA (Touchstone Applied Science Associates, Inc.) corpus [8] as a reference corpus for English writing samples. The corpus contains over 20,000 texts that contain at least two paragraphs from the following categories: business, health, home economics, industrial arts, language arts, science, social studies, miscellaneous and unmarked. The selection of minimum 2 paragraphs was necessary to compute the entire range of textual complexity indices that consider a document's paragraph structure. Table 1 presents the distribution of documents used in this preliminary analysis, for each category.

There are a few limitations to using TASA. First, the corpus does not represent student writing or essay writing, but rather sample writings from text books used in elementary and secondary educational settings. Second, the corpus contains repeated samples from some texts, which introduces repeated variance into statistical analyses. Nevertheless, TASA does provide a large sample of texts that appropriately represents the writing to which students are exposed and should accurately reproduce expected linguistic norms. Thus, for the pilot study reported below, the corpus provides an adequate representation of educational English.

Table 1. Documents in the dataset across all categories.

Category name	Essay count
Business	1,052
Health	507
Home Economics	43
Industrial Arts	64
Language Arts	9,537
Science	3,504
Social studies	6,945
Miscellaneous	744
Unmarked	1,168
Total	*23,564*

3.2 The *ReaderBench* Framework

This study is an extension of the *ReaderBench* framework [9], which already integrates various lexical, syntactic, semantic and discourse-centered complexity indices that were defined in previous studies. Indices computed at the surface level relate to lexical components of the text (word, sentence and paragraph counts, sentence lengths, commas and other punctuation marks) and fluency. Syntax measures refer to text analyses based on words and sentences, including part of speech tagging (POS) and corresponding syntactic dependencies. While adjectives and adverbs indicate a more detailed and complex text structure, pronouns denote a more linked discourse in terms of pronominal resolutions. Multiple indices are derived from the parsing tree which can provide valuable information - e.g., an increased number of dependencies indicate a complex discourse structure which leads to an increased textual complexity.

Lexical and semantic cohesion represent measures of a texts' organization from sentence level, to sentence transitions and paragraph interconnections [10]. A text with high cohesion has sentences linked in a logical manner. Within the *ReaderBench* framework, cohesion is computed by determining the semantic relatedness between various text components, such as words, sentences or paragraphs. *ReaderBench* makes extensive use of Cohesion Network Analysis (CNA) [11] which provides a comprehensive view of discourse by combining semantic cohesion with Social Network Analysis (SNA) measures applied on the cohesion graph (i.e., the internal representation of discourse structure). Semantic similarity [10] is measured in *ReaderBench* by considering both lexical proximity within the WordNet ontology (namely the Wu-Palmer semantic distance), as well as through the Latent Semantic Analysis (LSA), Latent Dirichlet Allocation (LDA) and word2vec semantic models. Besides CNA, *ReaderBench* is grounded in dialogism and implements a wide range of complexity indices related to inter-animation patterns of voices (i.e., points of view operationalized as semantic chains of related concepts) [12]. In addition, *ReaderBench* includes several vocabularies, for example the General Inquirer [13] that included the Lasswell dictionary.

3.3 Aggregating Complexity Indices into Principal Components

More than 740 complexity indices were generated through the *ReaderBench* framework, including all word-list indices. The high number of indices makes them difficult to interpret [4]. However, the linguistic features underlying student writing can likely be described through a composition of individual indices. Thus, we applied a Principal Component Analysis (PCA) [14] in order to group individual indices into components. Due to the high number of text indices and assumptions of PCA, various stages of index pruning were enforced, as follows.

Elimination of Localized Indices and of Outlier Documents

Indices that accounted for low linguistic coverage (more than 80% of the scores for all documents are 0 or −1, and thus irrelevant) were eliminated. In addition, we also removed documents for which at least 10% of indices are registered as outliers, namely we considered the value of an index for a document to be an outlier if it deviates more than 2 standard deviations from the mean value measured across the corpus. From the 745 complexity indices that the *ReaderBench* framework initially generated, 378 were removed by this step, with 367 indices being subjected to further pruning. In addition, a number of 3,052 documents were removed since they were marked as outliers for at least 10% of complexity indices.

Check for Normality

Indices that correspond to abnormal statistical distributions with respect to the skewness and kurtosis were removed as is required for PCA. With regards to skewness distribution, we removed all indices that have an absolute skewness value over 2. However, we found that there are few complexity indices with truly platykurtic distributions for the TASA dataset (i.e., a kurtosis value of no more than 3). Therefore, we have relaxed the threshold for which indices are eliminated by allowing any index that has a corresponding kurtosis value of at most 3.5. Out of the 367 complexity indices that passed the previous selection stage, 64 were found to have normal distribution.

Elimination of Strongly Correlated Indices

To further reduce the number of indices used in the PCA, a final pruning stage relying on the Pearson correlation coefficient (PCC) was enforced with the intention of removing redundant indices (those indices that strongly correlate to at least another index) which is an assumption of PCA. The PCC measures the linear correlation between two variables, giving values. If two indices have a strong positive or negative linear correlation, then they describe the dataset in a similar manner; as such, only one should be kept.

By applying a threshold over which an index pair is considered to be highly correlated, we removed indices, while minimizing information loss. This ensures that for each removed index, at least one index that was strongly correlated with it is still used by the PCA. Experimentally, we found that the best results are obtained when the threshold is set to .9, which means that any index pair with a corresponding absolute PCC of at least .9 is viewed as strongly correlated. The following heuristic greedy algorithm was applied on an undirected graph having as nodes the textual complexity indices, and edges reflecting high correlations between nodes. First, we sorted all node adjacency lists in decreasing order of their size - i.e., indices that have strong correlations with most other indices come first). Second, for each index we checked whether it has any redundancies (a non-empty adjacency list) at a given step. If the index has no redundancies, it was not eliminated. Otherwise, we checked if its elimination will cause any already-eliminated index to lose all redundancies (which would lead to information loss). If any such index exists, the previous index is kept; otherwise, the index is disregarded. By applying this approach, we reduced the list of textual complexity indices reported by *ReaderBench* to 45.

Principal Component Analysis (PCA)
After all previous pruning stages were complete, a PCA was performed on the remaining indices. The PCA was used to group similar indices into fewer, orthogonal components. In contrast, Factor Analysis (FA) relies on a formal model to predict observed variables from theoretical latent variables. In our case, the PCA was more appropriate as we are dealing with observable textual complexity indices and our aim is to identify comprehensible linear composites. The current version relies on R Engine (http://www.rforge.net/org/doc/org/rosuda/JRI/Rengine.html) to invoke from ReaderBench the PCA function *prcomp*. Results indicated that the optimum number of components was 5 (the explained variance was higher than .01 and the component contained at least two indices), with a total explained variance of .764. We then computed the loadings for the indices for these components. A heuristic approach was performed in order to determine which indices belong to which principal component. Namely, we set all loading factors under a certain threshold (in our experiments, the threshold value was set to .4) to be 0. If more than two components loaded in the same index with values over .4 (in absolute), we assigned the index to the component which has the largest eigenvector value for that index and set the rest to 0. Therefore, all the indices were uniquely assigned to a major component.

3.4   Comprehensive Feedback Generation Through an Extensible Rule-Based System

We designed and implemented an extensible rule-based engine for generating comprehensive user feedback. This system supports two types of rules: rules that use the above described components (for a component, the corresponding score is computed as a linear combination of the values of individual indices and their loading factors), and rules that only use basic index values (which are computed directly by the *ReaderBench* framework). For each rule, a minimum-maximum range, as well as corresponding feedback messages in case the document score exceeds the minimum or the maximum amount, are required. For the composite components, we computed the range using the average and standard deviation that is registered across all the documents from the TASA dataset. As such, for a given component, the accepted range is between the average component values across the corpus and ± 2 standard deviations for the value distributions. Other rules based on expert judgment (for example, the suggested number of paragraphs should be between 2 and 7) were also defined in order to provide more in-depth feedback.

Our system supports three levels of granularity targeted by the rules: document, paragraph and sentence levels. Each rule must define the level at which it is applied (for example, the following rule – "the number of words in a sentence should to be at least 10 and at most 35" – is applicable at sentence level). Multiple feedback messages can be defined for each of the two cases of each rule (namely if the measured value for a document is less than the minimum value or greater than the maximum value). For this study we have introduced on average 2 predefined feedback messages per rule that are interchangeable, but more can be easily added. Various predefined feedback messages provided to users help avoid monotony of feedback and potentially better understand the root cause, as well as enhancements.

3.5 A Multi-layered, Interactive Visualization Interface

A minimalistic visualization interface was designed in which users can input their essays and receive feedback at the previously defined granularity levels. For each level, the user can hover over any of the corresponding text segment (for example, at para- graph level the user can select each paragraph individually) to receive feedback for all rules triggered by the engine. Two color gradients were used to indicate severity. A gradient of red, with darker shades suggesting more issues with the passage, is used when at least one rule is triggered, with higher severities being assigned when the accepted intervals for rules are exceeded by multiple standard deviations. In contrast, a blue gradient is used for those textual segments for which no rules are triggered; darker shades suggest passages that are similar to the average values measured across the entire corpus. This allows an intuitive interface through which users can receive visual feedback for their essays.

4 Results

4.1 PCA Results

The PCA generated five principal components, each loading various individual *ReaderBench* indices. Each principal component explained a certain variance of the dataset, with the first component explaining the largest part of the total variance. Table 2 presents the corresponding component variances and the loading factors for each textual complexity index.

Interpreting the components and assigning them names requires an understanding of the indices that are being loaded, as well as knowledge of textual characteristics that may be exhibited by specific documents. We named the previous components as follows:

Table 2. Final loadings for the selected 5 components.

Index name (*Category from ReaderBench*)	V1	V2	V3	V4	V5
Average unique verbs per paragraph (*Syntax*)	−.836				
Average words in IAV word list per paragraph (*Vocabulary*)	−.832				
Average words in Active General Inquirer per paragraph (*Vocabulary*)	−.822				
Average unique prepositions per paragraph (*Syntax*)	−.808				
Average sentence linking connectors per paragraph (*Syntactic dependencies*)	−.771				
Average unique nouns per paragraph (*Syntax*)	−.753				
Average determiner syntactic dependencies per paragraph (*Syntactic dependencies*)	−.730				
Average nominal modifiers per paragraph (*Syntactic dependencies*)	−.727				
Average words in IAV word list per sentence (*Vocabulary*)	−.699				
Average syntactic dependencies per sentence (*Syntactic dependencies*)	−.695				
Average voice distribution per paragraph (*Dialogism*)	−.686				
Average words in Active General Inquirer per sentence (*Vocabulary*)	−.686				
Average number of verbs per sentence (*Syntax*)	−.683				
Average paragraph-document cohesion using Wu-Palmer semantic distance (*Cohesion*)	−.641				
Average case markings per sentence (*Syntactic dependencies*)	−.629				
Average determiner syntactic dependencies per sentence (*Syntactic dependencies*)	−.615				
Average unique adjectives per paragraph (*Syntax*)	−.614				
Average sentence length expressed as character count (*Surface*)	−.605				
Average inter-paragraph cohesion using word2vec (*Cohesion*)	−.447				
Average voice entropy at paragraph level (*Dialogism*)	.690				
Average sentence-paragraph cohesion using Wu-Palmer semantic distance (*Cohesion*)		−.893			
Average sentence-paragraph cohesion using LDA (*Cohesion*)		−.892			

(*continued*)

Table 2. (*continued*)

Index name (*Category from ReaderBench*)	V1	V2	V3	V4	V5
Average sentence-paragraph cohesion using LSA (*Cohesion*)		−.890			
Average sentence-paragraph cohesion using word2vec (*Cohesion*)		−.874			
Average sentence score from CNA (*Cohesion*)		−.486			
Average transition cohesion using Wu-Palmer semantic distance (*Cohesion*)			.492		
Average word depth hypernym tree (*Word complexity*)			.583		
Average inter-paragraph cohesion using Wu-Palmer semantic distance (*Cohesion*)			.648		
Average sentence voice cumulative effect (*Dialogism*)			.665		
Coverage percentage of lexical chains (*Dialogism*)			.774		
Average voice co-occurrence per paragraph (*Dialogism*)			.820		
Standard deviation of voice cumulative effect per paragraph (*Dialogism*)			.827		
Average voice cumulative effect per paragraph (*Dialogism*)			.865		
Average word polysemy count (*Word complexity*)				−.851	
Average word difference to corresponding stem as characters count (*Word complexity*)				.693	
Standard deviation of word characters (*Word complexity*)				.700	
Average word syllable count (*Word complexity*)				.841	
Average inter-paragraph cohesion using LSA (*Cohesion*)					−.804
Average transition cohesion using LSA (*Cohesion*)					−.737
Average inter-paragraph cohesion using LDA (*Cohesion*)					−.678
Average transition cohesion using LDA (*Cohesion*)					−.669
Average transition cohesion using word2vec (*Cohesion*)					−.621
Average start-end cohesion using LSA (*Cohesion*)					−.614
Explained Variance	**.395**	**.148**	**.124**	**.056**	**.041**

1. **V1**: *Readability* – this component inversely correlates with the increased usage of certain parts of speech (e.g., verbs, nouns, prepositions), discourse connections, increased number of different syntactic dependencies, voices' distribution at paragraph level, paragraph-document similarities, word statistics per paragraph and sentence levels using specific dictionaries (e.g., Active General Inquirer, IAV Laswell), and is directly proportional with the voice entropy (i.e., diversity) at paragraph level;

2. **V2**: *Local cohesion* – component which illustrates the semantic relatedness between sentences, namely sentence-paragraph cohesion computed using LSA, LDA, and the Wu-Palmer semantic distance, as well as the sentence scores determined using CNA;

3. **V3**: *Paragraph-level voice inter-animation* – this component refers to voices' inter-animation at paragraph level and it is directly proportional to voices' cumulative effect (both at paragraph and sentence levels), and voice co-occurrences within each paragraph;

4. **V4**: *Word complexity* – indicates the complexity of specific words expressed as an increase in word polysemy and syllable counts, as well as deviations of word characters from corresponding stems;

5. **V5**: *Global cohesion* – refers to the semantic relatedness between (inter) paragraphs and is inversely proportional with transition cohesion (semantic similarity between the last sentence of the previous paragraph and the first sentence of the following one) computed using LSA, LDA, and word2vec.

4.2   The Rule-Based Interface Behind *ReadME*

Figure 1 shows the results of the system' analysis at paragraph-level, with each paragraph being individually highlighted. The degree to which the system considers that a paragraph requires user's attention is highlighted in red. The middle paragraph is intuitively shown to have the most issues in terms of length (see Fig. 1 in which explanations are given in the left-hand side of the interface with regards to the issues detected in the second paragraph).

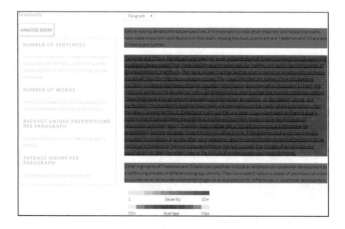

Fig. 1. Feedback provided at paragraph level.

By setting the granularity to sentence level, suggestions regarding individual sentences can be observed in Fig. 2. The sentence with highest number of issues is highlighted with a darker red shade, and the system suggests that the sentence should have its length reduced (in the left side). There are some sentences which are highlighted with blue gradients, meaning that they do not present abnormal values among the indices that are used by the sentence-level rules. The system intuitively suggests that the sentences with the darkest shade of blue require the least attention.

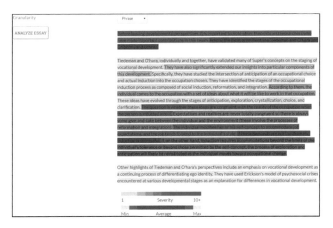

Fig. 2. Feedback provided at sentence level.

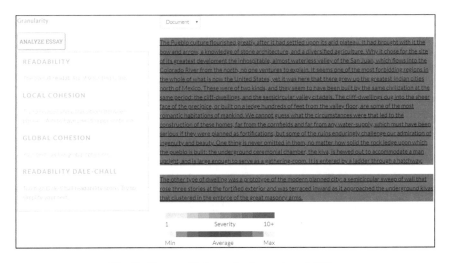

Fig. 3. Essay with low cohesion and readability.

In Fig. 3, a document-level view of a different essay is shown. The issues detected by the system suggest that the text has low readability and cohesiveness. Indeed, the depicted essay is highly technical, discussing the Pueblo culture's settlements in the

San Juan valley. The TASA dataset, as mentioned before, contains a varied collection of documents from various grade levels and subjects; it is thus expected that more technical texts such as this one will be considered less readable relative to TASA categories.

5   Conclusions

Measures of textual complexity can be used in order to provide both insights into numerical properties of various textual features, as well as text component characteristics such as readability, local or global cohesion, or word complexity. This paper presents an initial examination of our smart learning environment and reports on its ability to provide personalized recommendations to improve learner's writing style by aligning results reported by the tool to reference texts of various linguistic complexity.

Our method generates feedback using an extensible rule engine that contains multiple entries with requirements for certain linguistic components (i.e., complexity index or component) necessary for feedback generation. The minimalistic user-friendly web interface provides a convenient method to input an essay and receive direct visual cues with regards to linguistic features at various granularity levels (i.e., the entire document, a paragraph or a sentence). The color-based highlighting system combined with the textual suggestions work in tandem: the color system provides intuitive cues concerning any potential issues, whereas the text suggestions helping writers improve their work.

The strength of our automated writing evaluation method is dependent onto the effectiveness of the statistical methods used to express a text's quality or the writing proficiency exhibited in the training sample. Thus, a major limitation of the current version of *ReadME* is that it trained using texts meant for reading (the TASA dataset with its corresponding categories), which were afterwards used as baseline for writing tasks. Moreover, at present we did not perform any classroom experiments. We will continue to develop ReadME using specific training corpora that reflect different student writing levels and test our system with students of different levels. In addition, the visualization system can be further enhanced by introducing new granularities such as word-level feedback, as well as concurrent feedback for multiple documents by simultaneously comparing their document-level highlighting.

Acknowledgments. This research was partially supported by the README project "Interactive and Innovative application for evaluating the readability of texts in Romanian Language and for improving users' writing styles", contract no. 114/15.09.2017, MySMIS 2014 code 119286, as well as the FP7 2008-212578 LTfLL project.

References

1. Roscoe, R.D., Varner, L.K., Crossley, S.A., McNamara, D.S.: Developing pedagogically-guided algorithms for intelligent writing feedback. Int. J. Learn. Technol. **25**(8(4)), 362–381 (2013)
2. Attali, Y., Burstein, J.: Automated essay scoring with e-rater V.2.0. In: Annual Meeting of the International Association for Educational Assessment, p. 23. Association for Educational Assessment, Philadelphia (2004)
3. Warschauer, M., Ware, P.: Automated writing evaluation: defining the classroom research agenda. Lang. Teach. Res. **10**, 1–24 (2006)
4. Crossley, S.A., Kyle, K., McNamara, D.S.: To aggregate or not? Linguistic features in automatic essay scoring and feedback systems. J. Writ. Assess. **8**(1), 1–16 (2015)
5. McNamara, D.S., Crossley, S.A., Roscoe, R., Allen, L.K., Dai, J.: A hierarchical classification approach to automated essay scoring. Assess. Writ. **23**, 35–59 (2015)
6. Oelke, D., Spretke, D., Stoffel, A., Keim, D.A.: Visual readability analysis: how to make your writings easier to read. IEEE Trans. Visual Comput. Graph. **18**(5), 662–674 (2012)
7. McNamara, D.S., Graesser, A., Cai, Z., Dai, J.: Coh-Metrix Common Core TERA version 1.0, vol. 2018 (2013). http://coh-metrix.commoncoretera.com
8. Zeno, S.M., Ivens, S.H., Millard, R.T., Duvvuri, R.: The Educator's Word Frequency Guide. Touchstone Applied Science Associates Inc., Brewster (1995)
9. Dascalu, M., Stavarache, L.L., Dessus, P., Trausan-Matu, S., McNamara, D.S., Bianco, M.: ReaderBench: an integrated cohesion-centered framework. In: EC-TEL 2015, pp. 505–508. Springer, Toledo (2015)
10. Dascalu, M.: Analyzing discourse and text complexity for learning and collaborating. In: Studies in Computational Intelligence, vol. 534. Springer, Cham (2014)
11. Dascalu, M., McNamara, D.S., Trausan-Matu, S., Allen, L.K.: Cohesion network analysis of CSCL participation. Behav. Res. Methods, 1–16 (2017)
12. Dascalu, M., Allen, K.A., McNamara, D.S., Trausan-Matu, S., Crossley, S.A.: Modeling comprehension processes via automated analyses of dialogism. In: CogSci 2017, pp. 1884–1889. Cognitive Science Society, London (2017)
13. Stone, P.J., Dunphy, D.C., Smith, M.S.: The general inquirer: a computer approach to content analysis (1966)
14. Jolliffe, I.T.: Principal component analysis and factor analysis. In: Principal Component Analysis, pp. 115–128. Springer (1986)

What Is the Cat Doing? Supporting Adults in Using Interactive E-Books for Dialogic Reading

Stephanie Githa Nadarajah, Peder Walz Pedersen,
Camilla Gisela Hansen Schnatterbeck, Roman Arberg,
and Hendrik Knoche[(✉)] [iD]

Department of Architecture, Design and Media Technology, Aalborg University,
Rendsburggade 14, 9000 Aalborg, Denmark
stephanie.githa@gmail.com, pederstudio@gmail.com,
camilla.schnatterbeck@gmail.com, roman.arberg@gmail.com, hk@create.aau.dk

Abstract. Interactive e-books could provide a smart learning environment by providing adults with facilitation support for encouraging children to speak. To reap benefits for the adult reader and subsequently the children's language development, such support must be seamless to use for the reader and not impair the main experience of joint reading. This paper investigated, by means of video interaction analysis, how facilitation support can improve dialogic reading of daycare caregivers. Facilitation support consisted of providing good words and dialogic reading prompts on the top part of the screen outside the visual and textual story line of an interactive e-book. Ten caregivers with groups of two to three children between the ages of 22 and 48 months participated in the study. Caregivers in the facilitation support group used quality prompts more often than the control group. More prompting from caregivers correlated with more utterances from the children.

Keywords: Dialogic reading · Quality prompting
Interactive e-book · Reader support · Language development
Young children

1 Introduction

Young children with a limited vocabulary can have difficulties expressing themselves and therefore communicating. This can have long-term effects in their lives, due to early language development's substantial role in development of later language and literacy skills [13]. In order to resolve the problem of children with a limited vocabulary, pedagogical tools have been developed to be used with *Dialogic Reading*: An interactive shared book reading intervention in which the adult prompts and gives the child feedback, encouraging the child to become the storyteller. Using such tools requires adults to prepare by reading the book and related materials. This is not always realistic in daycare institutions or at homes where reading with the child may occur spontaneously. On the other hand,

© Springer International Publishing AG, part of Springer Nature 2019
H. Knoche et al. (Eds.): SLERD 2018, SIST 95, pp. 146–158, 2019.
https://doi.org/10.1007/978-3-319-92022-1_13

if the adult does not know how to perform dialogic reading, the child might not reap the all the benefits that dialogic reading holds.

To the best of our knowledge there has been no previous study on how to support adults in dialogic reading through an electronic interface such an interactive book. We use the terms facilitator and adult (e.g. caregiver or parent) interchangeably henceforth. Our study showed the beneficial effects of providing facilitation support for the adult through the user interface of an e-book. It contributes to the growing body of research on accelerating children's language development through shared-reading interventions. Such altered e-books provide a smart learning environment in which adults learn what options exist for dialogic reading and how to prompt more in a seamless way while children develop their language skills.

2   Background

Reading to children from an early age has long been considered an important step towards development of language and emergent literacy skills [3,4,8]. According to Elkin [9], children are active listeners and learners already from infancy. Around the age of six months, children undergo the necessary physical changes to engage in joint attention with another person, which allows for directing the attention of the child to e.g. an object or activity [21]. Shortly after, they can attend to the content of the book rather than its physical characteristics.

Despite research associating reading to infants with not only language and literacy development, but also other aspects of cognitive development, only few studies have considered book reading with young children less than 18 months [2,22]. An observational study by Senechal et al. on how parents read to infants (9, 17 and 27 months old) showed that parents used more prompting and feedback on older infants than younger infants for whom parents use elaborations and attention recruiting utterances [17]. Children, whose parents prompted and gave more feedback, made more utterances, suggesting that shared reading interventions that promote these behaviours in adults may be effective in fostering children's language development [17]. One such interactive shared reading intervention is dialogic reading.

Dialogic reading can be described as a process in which the adult presents new vocabulary and concepts through scaffolding to the child. Scaffolding through reading occurs when the adult supports the child in using the new vocabulary and concepts, such that the child can use the words in a context and relate the learned concepts to personal experiences. Dialogic reading has proven beneficial in fostering children's vocabulary, oral language complexity, narrative skills, and later language and literacy skills [7,10,20]. Using techniques like:

1. encouraging the child by asking questions or prompting,
2. expanding on the child's utterances,
3. praising the child's effort to tell the story,
4. recasting the story if it was unclear to the child, and
5. increasing the responsibility of the child to tell the story,

accelerated development of the child's language skills [6,10].

Regarding types of prompts, the literature contains a number of ways to encourage children to speak about or participate in a story at hand. For engaging and keeping the attention of young children (between four to 27 months) Honig & Shin suggested the use of **motoric prompts**, which encourage the child to make specific hand motions or imitate animal sounds. This was based on their their observational study in daycare centres in which reading to children of this age range only occurred for 1–2 min due to problems of keeping their attention [3]. Simiarly, Whitehurst and Zevenbergen suggested using simple **wh-prompts** (what, where, who, and why), e.g. "*What is this called?*" for young children (two to three years of age) and in first time readings when introducing new vocabulary [10]. After multiple readings the reader can use **open-ended prompts** to encourage children to use their own words to tell the story e.g. "*What do you see on this page?*" [10]. **Distancing prompts** encourage children to relate the content of the book to aspects of their own life (e.g. "*Do you have a pet?*"). Open-ended and distancing prompts encourage help children in becoming the story teller and practising language in the dialogic reading context [10,15].

However, when reading dialogically with groups of children, individual open-ended or distancing prompts can lead to situations in which some children become inactive. An advantage of using dialogic reading in small groups is that children can serve as each others role models [5]. According to Whitehurst and Zevenberg, the group context can induce difficulties for the adult, when having to adjust the level of prompts and feedback to the abilities of the children. The adult might have to adjust the interactions to the lowest common denominator of the group, which means that the children with more advanced abilities might not be challenged or the opposite, that interactions occur on a higher level than what is optimal for the individual child [7]. Whitehurst and Zevenberg recommended to engage in dialogic reading with individual children as much as possible [10]), but due to the caregiver-child ratio, which if often very low in daycare centres, reading with small groups of children is often practised [10,21].

More written text in picture books resulted in less opportunities for children to actively participate in the activity [7]. Further studies indicated that using books with social-emotional content had a positive effect on the child, since many of the perceptual and cognitive abilities were intimately connected with emotions (attention, learning, memory and decision-making) [5]. Children get the opportunity to be exposed to real-life content through social-emotional content, which in return creates an opportunity among participants to have conversations on, how to deal with specific real-life issues. If the adult takes this into account, when reading with small groups of children, the children get to practice knowledge concerning e.g. sharing and cooperating with each other using the book as a point of departure.

2.1   E-Book Reading

According to previous studies, e-books have the ability to promote word recognition, phonological awareness, and verbal knowledge in young children [14,19]. But, they may also divert children's attention from the content of the story, e.g. when interacting with multimedia features, which impairs story understanding and recall [12,14,15]. Electronic features (e.g. music, sound effects, and games) negatively affected dialogic reading and story comprehension of three and five-year-olds. Traditional books and e-books with electronic features turned off, fostered distancing and story-related problems. But, with electronic features turned on, speech of both parents and children became more behaviour-related (e.g. about the interaction with the book *"Don't touch that."*), rather than story-related. Regardless of the book type, children were able to recall characters and events. However, *"children who read traditional books were significantly better at remembering the content and sequence of events in a story than those who read books with electronic features"* [15]. According to Parish-Morris et al., mid-sentence pauses caused by children interacting with electronic features, might be the reason for the lower story comprehension, since this resulted in interference with the child's cognitive memory processes, making it difficult for the child to create a coherent presentation of the story line [15].

Opposed to electronic features that pause the story-line as in EC-books, activating the interactive elements (e.g. clickable hidden hotspots that appear on characters, objects and words appearing in the text) only after reading or listening to the story was featured in a study by Korat et al. [18]. The study focused on investigating word comprehension and phonological awareness among three to four years old children, when using an e-book versus a printed book in mother-child reading. Findings showed no difference between children in the two groups in terms of progress in language. However, the group with mothers who used e-books showed significantly higher tendency to, e.g., discuss the child's personal experiences and use distancing and discussing language to elaborate on story comprehension, which the authors suggested being due to the hotspots that supported the text content.

Several studies stressed the importance of congruence between media features in e-books and the story-line. For example, Labbo and Kuhn found that incongruence between electronic features and the storyline in CD-ROM story books caused negative effects on children's story comprehension [16]. The study was a qualitative study on a five-year-old child, who tried two different CD-ROM books. Incongruent features (e.g. animations not related to story) resulted in behaviours and comments from a low level of cognition (e.g. labelling). On the other hand, features congruent with the story resulted in affective appreciation by the child and higher-order cognitive processes, such as the child making predictions about the story and connecting to its own personal experiences [16].

Based on similar studies and research, Koratd and Shamir created an evaluation questionnaire for teachers, in order to help them decide whether an e-book supports the children's literacy development [1]. The authors emphasized the importance of reflecting on the child's age before choosing a book and recommended investigating, whether e.g. the characters, actions, amounts of text and narrative elements are appropriate for the child's age.

In summary, interactivity in e-books has the potential to promote and hinder aspects of a child's learning and development. This depends on several factors such as the media features in the e-book being congruent to the story and the e-book being age-appropriate. We only found one study that investigated the effects of multimedia features in dialogic reading, particularly in the case of an interactive e-book application for tablet devices and for young children [11]. No studies investigated how to support adults in dialogic reading in such settings. Twelve children in the age range from 16 to 33 months participated in a study by Knoche et al., in which three different caregivers facilitated a one-to-one interactive e-book application reading session. Increased interactions with multimedia features in the e-book did not reduce children's response to prompts or the length of the children's utterances and *"facilitation by different adults varied a lot both in terms of dialogic reading and how much they allowed or encouraged the children to interact with the application"* [11]. However, the study did not mention in detail, how the dialogic reading varied in terms of adults' prompting and feedback.

3   Study 1

We conducted an observational study at two Danish daycare centres, in order to investigate, how adults vary in terms of dialogic reading and to identify, which problems arise when using an interactive e-book.

Six female caregivers and fifteen children aged between 17 and 36 months participated in the study (8 males, 7 females). All of the caregivers and children were familiar with dialogic reading (except one caregiver) and participated in the study in familiar settings at daycare institutions. Caregivers facilitated reading sessions with small groups of children. Four caregivers facilitated sessions with groups of three the remaining two facilitated groups of two children. We used an interactive e-book obtained from a previous study [11], which covered two animal protagonists exploring the world. Caregivers were introduced to the e-book before the observation took place and the study was conducted in dedicated rooms at daycare centres. We used questionnaires, video cued recall sessions (where the caregivers watched the recorded activity with a test facilitator and described their experience), and interviews with the caregivers after the reading sessions to get insight into the problems that arouse during the activity and how they used dialogic reading.

We coded and transcribed the data from the observational study and looked into how the caregivers prompted, children responded and how much the participants interacted with the e-book. In the debrief, we asked the caregivers to elaborate on their experience.

3.1   Results and Discussion

Reading sessions lasted on average 10.8 min per facilitator (SD $= 2.4$ min) and caregivers prompted on average 3.45 times/minute (SD $= 0.15$ prompts/minute). Figure 1 shows the frequencies of prompts/minute across the different types of prompts. Wh-prompts occurred in all sessions and most often of all prompt types. Using wh-prompts are recommended with young children (2–3 years old) as mentioned previously and especially when reading the book with the child first time [10]. Particularly in the session, where the caregiver had a specialization in linguistic development and was familiar with dialogic reading, wh-prompts occurred most often, whereas in the session where the caregiver was not familiar with dialogic reading and without any language related specialization, wh-prompts occurred least.

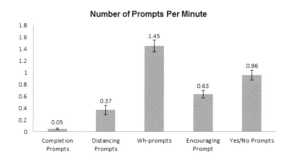

Fig. 1. Caregiver's number of prompts per minute across prompt types

None of the caregivers used open-ended prompts, which might be due to the reason that these types of prompts often require that children know the story beforehand. Multiple readings of the same book might make open-ended prompts more applicable for the facilitator, when trying to encourage the child to tell the story in his or her own words. Except completion prompts (fill-in-the-blank prompts, "*The lion roared and it said ...*") that only occurred 3 times, distancing prompts were the least used type of prompt. Results showed that two out of the six facilitators did not use distancing prompts at all. One of the cases was the caregiver, who was not familiar with dialogic reading. In the other case, 42% of the prompts the caregiver made, consisted of yes/no prompts, even though this caregiver in the interview mentioned reading dialogically with groups of children 2–3 times a day and being familiar with the technique. She used the yes/no prompts as commenting behaviour "*Is he sweet?*" after the child had already said "*He is sweet.*" and not in a labelling situation.

One of the caregivers mentioned in the video cued recall that she thought it was important to let the children talk about own personal experiences. However, the caregiver also mentioned that it was important to include both children and at the same time not hinder the first child in practising the language. When one of the children talked about flying and travelling, the caregiver had to take the control at some point and ask the other child in the activity whether she had tried flying before. Looking into distancing prompts in particular, it became evident that none of the distancing prompts used by the caregivers were related to social-emotional content, but rather specific to the child's personal experiences as e.g. which type of food they liked and where they went on vacation.

Encouraging prompts occurred in all sessions, too. This category included prompts that encouraged the child to use physical embodied interaction (e.g. "*Can you look up?*") or prompts related to interacting with the tablet device (e.g. "*Now you can try..*"). Investigating the interactions with the tablet, the caregivers had different opinions in terms of how much they should encourage the children. Two of the six caregivers mentioned that they found interactive elements in the e-book directly interrupting when reading. One of the caregivers elaborated during the video cued recall that this was due to many interactions being available in one scene. Another caregiver, whose children made most unrelated interactions (e.g. interactions that occurred during dialogue, where the child did not respond to the prompt from the adult, because the interactive elements captured the child's attention), mentioned that children should be allowed to explore. One facilitator commented on an interactive element in the application that was incongruent with the story-line. She pointed out that the interactive elements should rather support the story-line.

In summary, caregivers varied widely in how they prompted the children and thereby how they engaged the children to discuss the story content. We clearly saw the demands the caregivers faced during reading with groups of children in the targeted age. The caregivers had to be alert to a number of factors such as including all children and give the children enough time to respond. In addition to these demands some caregivers found that interactive elements in the e-book required additional attention during the activity at some point or that they interfered with the reading.

4   Study 2

In order to investigate the effects of supporting the adult through an interface, we designed and implemented an interactive e-book for tablets. The design was based on findings from previous work and our first study. It consisted of nine scenes/pages (see Fig. 2 for an example of a scene), each with a short story line sentence for the adult to read out loud (e.g. "*Tulle and Skralle live in this house.*"), interactive elements, multimedia features, a pause/play button and a facilitation support panel.

The pause/play button was a toggle button inspired by the previous study Korat et al. made [18]. It allowed the facilitator to control activation of interactive elements and multimedia features in the scene. Upon pressing the play button, interactive elements highlighted with an outline, animations and background sounds became activated and could be deactivated by pressing the button again. The features were deactivated by default. Interactive elements could be one of the main characters (see highlighted characters in Fig. 2). Pressing one of them triggered an animation (e.g. waving) and a sound effect (e.g. parrot greeting). We implemented only multimedia features congruent with the storyline and open-ended interactions in line with previous findings [11,16]. Because of the group context with multiple children there was no limit on how many times users could trigger the elements.

Facilitation support consisted of a list of good words and example prompts in the upper sixth of the screen. Each scene had between three and five good words (e.g. *Waving, Welcoming, Pets* and *Home* in example on Fig. 2), to inspire the facilitator to create prompt. Example prompts included simple what, where and why prompts (e.g. *"What animal is Tulle?"*, distancing (e.g. *"Do you have a pet?"*) and motoric (e.g. *"Can you wave?"*) prompts. It rotated through to a new set of prompts every five seconds. We omitted distancing prompts due to the target age range but included prompts that meant to encourage the discussion of social-emotional content (e.g. *"Have you been sad before?"*).

Participants. Eleven female caregivers from six different Danish daycare institutions participated in the study. A demographic questionnaire queried age, gender, educational background, prior experience with dialogic reading, mobile/tablet devices and familiarity with the content of the book. Caregivers were randomly assigned to either treatment group or control group. Five of the eleven caregivers were placed in a control group and of the remaining six caregivers, only five were included in the analysis. This was due to one of the caregivers in the treatment group not having the same foundation as the other caregivers to conduct the reading session (e.g. difficulties with basic interactions on the tablet and did not go over the application before the reading session as the other caregivers). Four out of five facilitators were familiar with dialogic

Fig. 2. Screenshot of the application with facilitation support that consisted of good words and prompts placed in the upper panel of the user interface

reading in the control group (1 used it often, 2 used it sometimes, 1 used it rarely), whereas all facilitators were familiar with dialogic reading in the treatment group (4 used it often, 1 used it sometimes).

Twenty-five children (12 females, 13 males) aged between 22 and 48 months participated (mean $= 30.1$, SD $= 6.76$) in the study. Children in the control group (N $= 14$) were 28 months old on average (SD $= 4.7$), while the average in the treatment group (N $= 11$) was 33 months (SD $= 8.12$). Four caregivers facilitated groups of three children and one caregiver facilitated a group of two children in the control group (mean $= 2.8$ children), while four caregivers facilitated groups of two children and one caregiver facilitated a group of three children in the treatment group (mean $= 2.2$ children).

Materials and Procedure. The treatment group used the application with facilitation support, while an identical application without the support was given to the control group. The reading sessions were audio- and video-recorded. All of the caregivers were allowed to go through the application before the reading session with the children. The treatment group went through a tutorial in the interface demonstrating, how to use the facilitation support. The observation took place in dedicated rooms at the daycare centres in settings familiar to the caregivers and the children. After the reading session, questionnaires and interviews were conducted with the caregivers.

We decided to only log the first 10 min of each session, due to possible difficulties keeping the attention of the child, according to caregivers at the daycare institutions. Some sessions took less than 10 min and resulted in 'free play', which was not included in the analysis. Prompts made by facilitators were either categorized as wh-prompts, distancing prompts, motoric prompts, encouraging prompts or yes/no prompts. The first three prompt types are henceforth also referred to as *quality prompts*. Counts of prompts were normalized in relation to time to get prompts/minute and marked with a 'yes' or 'no', if they were potentially used or inspired by the facilitation support. Children's utterances were categorized as either short (e.g. "*yes*", "*no*", "*hmm*"), story-related, incomprehensible, or other, if unrelated to the story context.

4.1   Results

According to a Mann-Whitney U test caregivers prompted significantly more often ($p < .01$) in the treatment group (5.5 prompts/minute) than in the control group (2.1 prompts/minute). A multiple regression analysis showed that neither *group size* ($p = .08$), nor children's *age* ($p = .73$), nor the caregivers previous *dialogic reading experience* ($p = .11$) were significant predictors of prompting frequency.

We investigated the use of wh-, distancing, and motoric prompts separately (see Table 1). Caregivers in the treatment group used distancing prompts significantly more than the caregivers in the control group. Even though the treatment group also prompted more wh- and motoric prompts, these differences were not significant.

Table 1. Types of caregivers prompts per minute in the control and treatment group

Group	Wh- $p = .09$	Distancing $p < .01$	Motoric $p = .14$
Control	1.6	0.1	0.5
Treatment	2.6	2.0	0.9

Some caregivers used the facilitation support provided in the user interface more than others. One caregiver was only inspired by or potentially used the support in 14% of her prompts, while another was inspired by or directly used in 89% of her prompts. On average, 58% of the used prompts were inspired by or potentially used from the facilitation support (SD $= 28.67\%$).

Children in the treatment group uttered more words (12.98 words/minute) than children in the control group (3.15 words/minute) significantly different according to a Mann-Whitney U test ($p < .01$). We controlled for some of the variables that might have had an impact on the results, which included the caregivers previous *dialogic reading experience*, children's *age*, *group size*, and *prompting frequency*. Using regression we found no significant relationship between how often the caregivers had performed dialogic reading before the study and the number of words per minute during the study ($p = .42$). However, we did find a significant relationship between age and words/minute according to a correlation test ($r = 0.56$, $p < .01$). Knowing that words/minute depended on the child's age, we included age as a predictor in a multiple regression analysis investigating, if children in general uttered more, when being exposed to more wh- and distancing prompts/minute. Motoric prompts were not included due to the fact that it supported physical activity rather than verbal activity. According to the regression, children uttered significantly more when exposed to a higher frequency of prompts ($p = .04$). Figure 3 plots the children's words/minute against children's age (in months) and prompts/minute.

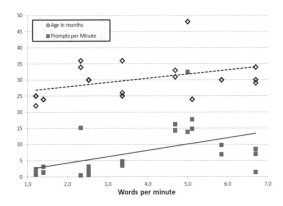

Fig. 3. Relationship between children's words/minute plotted against children's age (in months) and quality prompts/minute

The mean age of the children in the treatment group was higher than in the control group, but even though the frequency of quality prompts were significantly higher in the treatment group, a multiple regression analysis did not support that children uttered significantly more, having age and prompts/minute as predictors. We used an ANOVA test to investigate, whether group size had an impact on the children's utterances. Results showed that children who were in groups of two uttered significantly more than children in groups of three $(p < .01)$. The result of this can maybe be explained by the fact that groups of two were exposed to more personal prompts compared to the three-man groups, due to an unbalanced design. Using age, prompting frequency, and group size as predictors in a multiple regression analysis to predict children's utterances, only group size showed having a significant effect $(p < .01)$.

According to the interviews, caregivers had different opinions on the facilitation support and how they used it. All facilitators in the treatment group agreed on the fact that the facilitation support enhanced the dialogue and did not distract the children. However, one facilitator stated that she did not use it very often, but would recommend it as a feature for facilitators without any knowledge of dialogic reading. Some of the facilitators thought some prompts were a bit too difficult for the children and that some prompts would be more appropriate after repeated readings with the children. In the questionnaire, we found that some facilitators stated having used the good words more, while others used the prompts more often. One of the caregivers who prior to this study had used a piece of paper with supporting prompts during dialogic reading mentioned: "*I thought it [facilitation support] worked really well. Usually we sit next to a paper to read with, but here it is available all the time, so I can prepare questions while the children are talking*".

4.2   Discussion

The caregivers in the treatment group prompted more frequently and used more quality prompting than those in the control group. Higher prompting frequencies resulted in more speaking from the children and there were no other significant differences between treatment and control. Children in the treatment group uttered significantly more than children in the control group. The age of the children and group size impacted on children's utterances, in line with previous research. Whitehurst & Zevenbergen in particular, mentioned that group size was a variable that had an effect on how much all children were included and thereby how much they uttered in a reading session.

Even though caregivers prompted more in the treatment group and might have used the facilitation support to do so, we stress the fact that our design does not take into account the abilities of the children. The caregivers still had to adjust their questions and feedback to the children's level. Prompting constitutes only one aspect of dialogic reading and further studies should investigate other strategies, such as expanding the children answers, repetition, and presenting new words through scaffolding. These strategies may not come naturally for people without or low experience in dialogic reading and may require additional support.

Our study had a few other limitations. We observed several times that some children became inactive as a result of the adult asking a question directed at one child. Our study did not control for such behaviour. Investigating the verbal activity of the children, we only looked into number of utterances and not whether utterances happened as a response to a prompt. Another important factor for children utterances is that, in order to respond to the adult's prompts in dialogic reading, children receive scaffolding and get encouraged throughout multiple reading sessions, such that they become the story tellers in the end. We only investigated the first session of the book reading in which the children respond least. All facilitators in the treatment group were familiar with dialogic reading and practised it often with children in our study.

5 Conclusion

This study addresses the potentials in utilizing user interfaces in e-books to support the adult. Our findings suggest that providing the facilitator with good words and prompts through the interface supported the facilitator in using more and higher quality prompts. Whether this has an effect on children's utterances was discussed in this study. Future studies should investigate the actual learning benefits for children, when providing facilitation support through the interface both short and long term and how such learning environments can be made smarter for all involved stakeholders.

Acknowledgments. We would like to thank Ditte Aarup Johnsen for the collaboration and permission to use her story-line and artwork of Tulle and Skralle. Thanks to the caregivers, children at the Bornholmsgade day care center, and their parents whose consent and involvement enabled the conduct of this research.

References

1. Adina Shamir, O.K.: How to select CD-ROM storybooks for young children - the teacher's role. Read. Teach. **59**(6), 532–543 (2006)
2. Murray, A., Egan, S.M.: Does reading to infants benefit their cognitive development at 9-months-old? An investigation using a large birth cohort survey. Child Lang. Teach. Ther. **30**(3), 303–315 (2014). https://doi.org/10.1177/0265659013513813
3. Honig, A.S., Shin, M.: Reading aloud with infants and toddlers in child care settings: an observational study. Early Child. Educ. J. **28**(3), 193–197 (2001). https://doi.org/10.1023/A:1026551403754
4. Kupetz, B.N., Green, E.J.: Sharing books with infants and toddlers: facing the challenges. Young Child. **52**(2), 22–27 (1997)
5. Doyle, B.G., Bramwell, W.: Promoting emergent literacy and social emotional learning through dialogic reading. Read. Teach. **59**(6), 554–564 (2006). https://doi.org/10.1598/RT.59.6.5
6. Trivette, C.M., Dunst, C.J.: Relative effectiveness of dialogic, interactive, and shared reading interventions. CELLReviews **1**(2), 1–11 (2007)

7. Lonigan, C.J., Whitehurst, G.J.: Relative efficacy of parent and teacher involvement in a shared-reading intervention for preschol children from low-income backgrounds. Early Child. Res. Q. **13**(2), 263–290 (1998). https://doi.org/10.1016/S0885-2006(99)80038-6

8. DeBaryshe, B.D.: Maternal belief systems: linchpin in the home reading process. J. Appl. Dev. Psychol. **16**(1), 1–20 (1995). https://doi.org/10.1016/0193-3973(95)90013-6

9. Elkin, J.: Babies need books in the critical early years of life. New Rev. Child. Lit. Librariansh. **20**(1), 40–63 (2014). https://doi.org/10.1080/13614541.2014.863666

10. Whitehurst, G.J., Zevenbergen, A.A.: Dialogic reading: a shared picture book reading intervention for preschoolers. In: On Reading Books to Children, pp. 177–200 (2003)

11. Knoche, H., Rasmussen, N.A., Boldreel, K., et al.: Do interactions speak louder than words? Dialogic reading of an interactive tablet-based e-book with children between 16 months and three years of age. In: Proceedings of the 2014 Conference on Interaction Design and Children, IDC 2014, pp. 285–288 (2014). https://doi.org/10.1145/2593968.2610473

12. John Trushell, A.M.: Primary pupils recall of interactive storybooks on CD-ROM: inconsiderate interactive features and forgetting. Br. J. Educ. Technol. **36**(1), 57–66 (2005). https://doi.org/10.1111/j.1467-8535.2005.00438.x

13. Preston, J.L., Frost, S.J., Mencl, W.E.: Early and late talkers: school-age language, literacy and neurolinguistic differences. Brain **133**(8), 453–469 (2010). https://doi.org/10.1177/089443939201000402

14. de Maria, T., Jong, A.G.B.: Quality of book-reading matters for emergent readers: an experiment with the same book in a regular or electronic format. J. Educ. Psychol. **94**(1), 145–155 (2002). https://doi.org/10.1037//0022-0663.94.1.145

15. Parish-Morris, J., Mahajan, N., Hirsh-Pasek, K., et al.: Once upon a time: parent-child dialogue and storybook reading in the electronic era. Mind Brain Educ. **7**(3), 200–211 (2013). https://doi.org/10.1111/mbe.12028

16. Labbo, L.D., Kuhn, M.R.: Weaving chains of affect and cognition - a young child's understanding of CD-ROM talking books. J. Lit. Res. **32**(2), 187–210 (2000). https://doi.org/10.1080/10862960009548073

17. Senechal, M., Cornell, E.H., Broda, L.S.: Age-related differences in the organization of parent-infant interactions during picture-book reading. Early Child. Res. Q. **10**(3), 317–337 (1995). https://doi.org/10.1016/0885-2006(95)90010-1

18. Korat, O., Shamir, A., Heibal, S.: Expanding the boundaries of shared book reading: E-books and printed books in parent-child reading as support for children's language. First Lang. **33**(5), 504–523 (2013). https://doi.org/10.1177/0142723713503148

19. Pav Chera, C.W.: Animated multimedia 'talking books' can promote phonological awareness in children beginning to read. Learn. Instr. **13**(1), 33–52 (2006). https://doi.org/10.1016/S0959-4752(01)00035-4

20. Ping, M.T.: Group interactions in dialogic book reading activities as a language learning context in preschool. Learn. Cult. Soc. Interact. **3**(2), 146–158 (2014). https://doi.org/10.1016/j.lcsi.2014.03.001

21. Soundy, C.S.: Nurturing literacy with infants and toddlers in group settings. Child. Educ. **73**(3), 149–153 (2012). https://doi.org/10.1080/00094056.1997.10522673

22. Richman, W.A., Colombo, J.: Joint book reading in the second year and vocabulary outcomes. J. Res. Child. Educ. **21**(3), 242–253 (2007). https://doi.org/10.1080/02568540709594592

The Robbers and the Others – A Serious Game Using Natural Language Processing

Irina Toma[1], Stefan Mihai Brighiu[1], Mihai Dascalu[1,2(✉)], and Stefan Trausan-Matu[1,2]

[1] University Politehnica of Bucharest, 313 Splaiul Independentei, 060042 Bucharest, Romania
irina_toma@rocketmail.com, sbrighiu@gmail.com,
{mihai.dascalu,stefan.trausan}@cs.pub.ro
[2] Academy of Romanian Scientists, 54 Splaiul Independenţei, 050094 Bucharest, Romania

Abstract. Learning a new language includes multiple aspects, from vocabulary acquisition to exercising words in sentences, and developing discourse building capabilities. In most learning scenarios, students learn individually and interact only during classes; therefore, it is difficult to enhance their communication and collaboration skills. The prototype game described in this paper aims to fill this gap and improve the students' learning skills in a smart learning environment suitable for a problem-solving game. In addition, the game is also an useful tool for teachers because of the integrated chat analysis that enables the identification of the most predominant points of view and the overall level of collaboration between participants. The game is developed as a bot on the Slack chat platform and reacts to user commands. At the end of a game round, the bot is to save the conversation transcript and send it to the *ReaderBench* framework for further chat analysis.

Keywords: Collaborative serious games · Smart learning environment
Conversation analysis · Online chat · Language learning
ReaderBench framework

1 Motivation

One important task in a 'smart' learning ecosystem is to help language learning both native children and foreigners that enter in that ecosystem. However, language learning is a long-term process that includes vocabulary acquisition, accompanied by its usage in a coherent manner, within more and more elaborated contexts. The language learning process that is currently available in schools and universities is based on memorizing concepts, repeating them and using them in sentences [1]. Our study focuses on enhancing these methods by incorporating them in a smart learning environment that makes use of gamification. Starting from serious games centered on vocabulary acquisition, such as Semantic Boggle [2] and Semantic Taboo [3], our aim is to move forward towards evaluating how students understand written texts [4].

In this paper we introduce a prototype serious game that focuses on improving the collaboration and communication levels of students. The game, *The Robbers and the Others*, is a problem-solving role-playing smart chat environment in which students are

© Springer International Publishing AG, part of Springer Nature 2019
H. Knoche et al. (Eds.): SLERD 2018, SIST 95, pp. 159–164, 2019.
https://doi.org/10.1007/978-3-319-92022-1_14

encouraged to discuss in order to find the people responsible for robbing a bank. The game is presented as a Slack bot [5], and uses ElasticSearch [6] and Firebase (https://firebase.google.com/) to store messages and search through the conversations.

2 Chatbots

Chatbots are software programs that interact with humans using a Natural Language Processing (NLP) interface. The concept emerged in the 1960s when the chatbot implementation was straightforward: it searched for simple keywords that matched the input of the user [7]. Nowadays, chatbots evolved into intelligent systems, which are no longer limited to pattern-matching or decision trees, but rely on NLP techniques and Machine Learning [8]. These systems are found in everyday life: education system, public relations sector, enterprise communication, and entertainment [9].

In education, chatbots have been used as Intelligent Tutoring Systems [10, 11] to help students learn new concepts. For example, *Chatbot* [12] is a didactic software instrument built to help students learn Computer Science concepts by programming the bot. The bot can connect to social media chats and reply to conversations. It provides contextual answers depending on the previous topics or in relation to the inputs of the other chat participants. Chatbot is given pairs of pattern-effect elements and responds with the effect when the pattern is matched. Students are responsible of writing these patterns and, by doing this, they learn basic Computer Science concepts.

In the enterprise communication area, many new chat systems emerge (e.g., Slack – www.slack.com, Skype for Business – www.skype.com/en/business, HipChat – www.hipchat.com). These integrate bots from the simplest ones, for tasks such as file sharing, notifications or white board for drawing schemas, to more complex ones, like triggering commands from textual conversations or agile management.

For the current work, Slack bots are of interest. Slack offers an API for creating bots that act similar to a human user. Some examples of bots include:

- *Sofi* (www.swipesapp.com) is a project manager chatbot available for Slack that assigns tasks to other people in the team at the manager's command. *Sofi* also chats with people about those tasks and when they are done, it assigns the next one.
- *MeetingBoot* (https://meetingbot.io/) checks availability of other people, searches for the best meeting times, books meeting rooms and notifies people when they are late for a meeting.

3 Game Description

The Robbers and the Others is a collaborative problem-solving game to be played in a group of minimum seven people. The purpose of the game is to solve a bank robbery case, identify the robbers and send them to jail. All interactions take place using text contributions and no images or animations are displayed to the users, only narrative segments. Moreover, all witness statements are collected through a Slack chat, enabling follow-up in-depth analyses of the transcripts.

The game exposes three types of users: (a) admins – they can start/stop sessions and can add/remove other admins; (b) leader robber and others – the leader plans the robbery, while the others have access to the robbery details and the list of robbers; (c) detective lead and others – the leader setups the press release, while others have access to the press release details, the detective list, and the topics. The users interact in the chat window during the stages of the game, each described in detail below.

Robbery and Press Release Scenes. The plot of the game takes place in a fictional town where the bank is robbed. The robbing involves shooting and the police arrives too late; thus, specifications are unclear and witness questioning is required.

The actual game is started by the admin once all required users joined the chat room. In the first step, the robbers and the detectives are chosen randomly. For a minimum of seven players, two of them will be robbers, two detectives and the rest civilians. Each of them receives a private message similar to the one in Fig. 1. Robbers and detectives should discuss in different chat rooms the details of the robbery and, respectively, of the press release (see Fig. 2a).

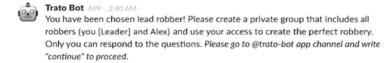

Fig. 1. The message sent to the leader robber at start up.

(a)   (b)

Fig. 2. The lead robber/detective decides the details of the robbery/press release.

After investigating the robbery scene and discussing with the witnesses over the chat, detectives discover information about the robbery: number of casualties, the weapons used by the robbers, and how much money was stolen. Next, they organize a press release that is not attended by the robbers.

For the press release, detectives can decide to change the information about the robbery in order to confuse the thieves. For example, if the robbery took place on a Saturday, they could say it happened on Sunday, or they could change the amount of stolen money (see Fig. 2b).

Town Hall Meeting. The plot twist of the story happens when the detectives announce the concerned citizens that they found the leader of the robbers and they are confident they would catch him soon. Hearing this news about the unknown leader, the robbers get frightened. During the following week, the detectives organize a new meeting in which the entire community presence is requested, including the robbers. The purpose of this meeting is to set some ground rules for the next discussions and decide on the list of topics. Detectives are in charge of changing the topic of the conversation, as some topics could reveal valuable information about the robbery. Each conversation is recorded by the Court Reporter and is available later on as XML transcripts. Figure 3a presents an excerpt from the conversation on the topic "Amount stolen". Here, the lead robber makes the first mistake, acting surprised when told about the amount of stolen money.

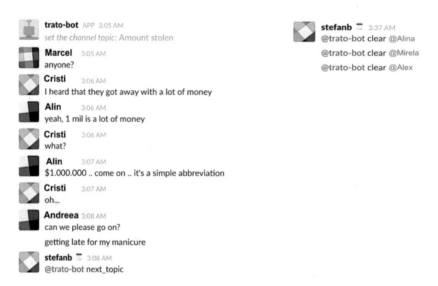

Fig. 3. **(a)** The lead robber makes the first mistake. **(b)** The detective clears two citizens and one robber.

Game End. After the topics end, the detectives lock people in the town hall for few minutes; the Court Reporter also leaves, so the discussions are not recorded. The robbers panic when the detectives return and say they solved the case. Detectives have no evidence to convict anyone, but they hope that the robbers will confess if put under pressure. They call the names of most people in the room, but, one by one, people are dismissed as innocent. The rest of the citizens would be held for additional questioning. As a final push, after 24 h, detectives decide to lower the robbers' sentence if they confess. Depending on the robbers, they would all get away if nobody confessed, or go to jail otherwise. The game can be lost by the detectives if they clear more robbers than 50% plus one robber, or won if the number of cleared civilians is higher than 50% minus one robber. As seen in Fig. 3b, the lead detective clears one of the robbers; therefore, the robbers win.

4   Chat Analysis

At any point in the game, the admin user can request the chat transcript. Once the transcript is generated, the bot sends it to the Computer Supported Collaborative Learning processing endpoint from the *ReaderBench* framework [13] for analysis. The response covers the level of involvement in the chat corresponding to each student. This information is most valuable in a visual format, where teachers can follow the contributions of each student, their evolution in time in terms of active participation (see Fig. 4), or of collaboration with their peers (see Fig. 5). Based on this information, teachers can decide how students collaborated in solving the problem, whether they understood the discussed topics and propose exercises to improve students' collaboration and comprehension levels.

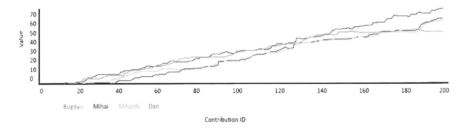

Fig. 4. Participant evolution graph.

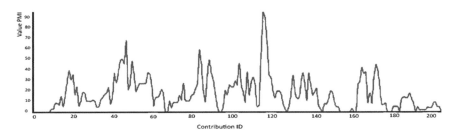

Fig. 5. Collaboration between participants throughout the conversation.

5   Conclusion

On the path to language learning, students are first confronted with vocabulary acquisition. Having a strong vocabulary base, they feel confident in moving towards the next step – understanding written texts and participating in conversations. Our study focuses on providing students an alternative language learning process, a smart gamified environment in which they learn by playing. The serious game prototype presented in this paper focuses on improving students' communication and collaboration skills. This is achieved in a fun, competitive, and smart environment where students play the roles of thieves, detectives, and civilians in a bank robbery. All student interactions are held in a chat environment, provided by Slack, where the conversation flow is dictated by a Slack bot called *trato-bot*. The educational value of the game comes from the integrated in-depth chat analyses based on the conversation transcripts. This is valuable for teachers as they can easily identify the most active students in the game, provide them feedback and suggest exercises to improve students' comprehension and collaboration levels.

Acknowledgements. This research was partially supported by the 644187 EC H2020 *Realising an Applied Gaming Eco-system* (RAGE) project and by the FP7 2008-212578 LTfLL project.

References

1. Thornbury, S.: How to teach vocabulary, vol. 1. Pearson Education India, Longman Essex (2006)
2. Toma, I., Alexandru, C.-E., Dascalu, M., Dessus, P., Trausan-Matu, S.: Semantic boggle: a game for vocabulary acquisition. In: European Conference on Technology Enhanced Learning, pp. 606–609. Springer (2017)
3. Toma, I., Alexandru, C.-E., Dascalu, M., Dessus, P., Trausan-Matu, S.: Semantic taboo–a serious game for vocabulary acquisition. Rom. J. Hum.-Comput. Interact. **10**(2), 241–256 (2017)
4. Toma, I., Iustinian, A., Dascalu, M., Trausan-Matu, S.: Reading space secrets-a serious game centered on reading strategies. Rom. J. Hum.-Comput. Interact. **9**(4), 269 (2016)
5. Shevat, A.: Designing Bots: Creating Conversational Experiences. O'Reilly Media, Inc. (2017)
6. Gormley, C., Tong, Z.: Elasticsearch: The Definitive Guide: A Distributed Real-Time Search and Analytics Engine. O'Reilly Media, Inc. (2015)
7. Shawar, B.A., Atwell, E.: Chatbots: are they really useful? Ldv Forum **22**, 29–49 (2007)
8. Schuetzler, R.M., Grimes, M., Giboney, J.S., Buckman, J.: Facilitating natural conversational agent interactions: lessons from a deception experiment (2014)
9. Dale, R.: The return of the chatbots. Natural Lang. Eng. **22**(5), 811–817 (2016)
10. Graesser, A.C., Chipman, P., Haynes, B.C., Olney, A.: AutoTutor: An intelligent tutoring system with mixed-initiative dialogue. IEEE Trans. Educ. **48**(4), 612–618 (2005)
11. Hämäläinen, W., Vinni, M.: Comparison of machine learning methods for intelligent tutoring systems. In: International Conference in Intelligent Tutoring Systems, pp. 525–534. Springer, Jhongli (2006)
12. Benotti, L., Martínez, M.C., Schapachnik, F.: Engaging high school students using chatbots. In: Proceedings of the 2014 Conference on Innovation & Technology in Computer Science Education, pp. 63–68. ACM (2014)
13. Gutu, G., Dascalu, M., Trausan-Matu, S., Dessus, P.: ReaderBench goes online: a comprehension-centered framework for educational purposes. In: RoCHI 2016, pp. 95–102. MATRIX ROM, Iasi (2016)

Modelling Smart Learning

Identifying Students Struggling in Courses by Analyzing Exam Grades, Self-reported Measures and Study Activities

Bianca Clavio Christensen[✉], Brian Bemman, Hendrik Knoche, and Rikke Gade

Aalborg University, Rendsburggade 14, 9000 Aalborg, Denmark
{bcch,bb,hk,rg}@create.aau.dk

Abstract. Technical educations often experience poor student performance and consequently high rates of attrition. Providing students with early feedback on their learning progress can assist students in self-study activities or in their decision-making process regarding a change in educational direction. In this paper, we present a set of instruments designed to identify at-risk undergraduate students in a Problem-based Learning (PBL) university, using an introductory programming course between two campus locations as a case study. Collectively, these instruments form the basis of a proposed learning ecosystem designed to identify struggling students by predicting their final exam grades in this course. We implemented this ecosystem at one of the two campus locations and analyzed how well the obtained data predicted the final exam grades compared to the other campus, where midterm exam grades alone were used in the prediction model. Results of a multiple linear regression model found several significant assessment predictors related to how often students attempted self-guided course assignments and their self-reported programming experience, among others.

Keywords: Academic performance · Student retention
Learning Management System · Learning Tools Interoperability
Problem-based Learning · Flipped learning

1 Introduction

Students enrolled in educations with technical content often struggle with passing technical courses and frequently drop out as a result [4,9]. Much of the research on student dropouts or retention has focused on the personality traits of a student, typically without also considering their actual learning progress. Both struggling students and course instructors, however, can benefit from an understanding of how the learning process of students is progressing. Such an understanding, for example, might encourage students to engage more deeply

© Springer International Publishing AG, part of Springer Nature 2019
H. Knoche et al. (Eds.): SLERD 2018, SIST 95, pp. 167–176, 2019.
https://doi.org/10.1007/978-3-319-92022-1_15

with the learning material or allow instructors to better direct resources to those in need. With many openly accessible learning resources, such as *Massive Open Online Courses (MOOCs)* now available, teachers are able to construct diverse learning ecosystems for their students which extend far beyond institutionally managed, digital *Learning Management Systems (LMSs)*. The research question addressed here is how information gathered from the diverse interactions students have with these learning resources can be used to identify struggling individuals.

In this paper, we present a set of instruments designed to identify struggling first-semester undergraduate students enrolled in an introductory programming course at a *Problem-based Learning (PBL)* university in Denmark, named Aalborg University (AAU). These instruments consist of both student self-reported personal attributes and self-assessed measures of the learning progress. We use these instruments in the construction of a multiple linear regression model for predicting student final exam scores. Our proposed model consists of significant predictors from this set of instruments that suggest a possible relationship between the academic success of a student and select personal attributes and learning progress measures. Importantly, these predictors could provide the university with the means to more effectively identify struggling students who may be at risk of leaving the education, allowing them to offer guidance to these individuals as early in their education as possible.

2   Background

Previous research on student retention has identified a number of factors for decreasing the risk of students leaving educational programs: *growth mindset* [11], *grit* (i.e., perseverance when faced with challenges) [10], *study habits* [27], *high school habits* [15,19], and *social support for studying* [5]. Although this research has documented a wide range of potential predictors of student retention, agreement between studies is low [18,20,26]. For this reason, continued research would be better served by considering case studies [9]. This could be done, for example, by detecting students at risk of leaving the education and then directing adequate resources to those individuals based on relevant features of the study program from which these students left. One previous study on student dropouts [4], looked at first semester students in an undergraduate Media Technology (MEA) program at AAU. These findings provided some evidence, in the form of interviews and study diary logs, that suggested that the skills in mathematics and programming required of the program were higher than students initially expected, resulting in a high number of dropouts. Natural science courses, for example, are notorious for low retention and first-time success rates, particularly in the first year of study [25]. It is essential then to investigate interdisciplinary educations, such as MEA, that combine technical, scientific, and design skills.

Engaging students in the learning process and making them responsible for their own progress is one of the primary goals of university education and PBL, in particular. One important reason for this is the comparatively less interaction and feedback students receive at a university than they are used to in high school. The principle of *pre-training* [23] suggests that cognitive overload can be reduced by providing students with basic information ahead of their actual lectures. This principle is often implemented online as self-study activities with *flipped learning*. Such an approach leaves more time for the instructor to facilitate classroom activities that are essential to a PBL framework, where students are required to analyze, evaluate, and create content in a hands-on fashion [17]. These activities in PBL can include scaffolding more complex concepts and skills through interaction, group work, peer feedback, and immediate teacher support [16]. This stands in contrast to, for example, the earlier stages in Bloom's taxonomy of learning (remember, understand, and apply) [3].

Breaking material down into smaller parts is one way to reduce the cognitive load of students [1] and doing so can make that content more accessible, focused, and easier to digest. Self-assessment questions are one way to increase this sort of in-depth learning in students [12], e.g., when a student is trying to understand where they went wrong on a quiz. Through self-assessment questions, instructors can efficiently assess and manage student learning by creating, for example, online assignments with automated grading and feedback. This is a fundamental approach advocated in *Learning Tools Interoperability (LTI)* [21]. Moreover, self-assessment quizzes of this type can be designed to adapt and grow in response to student performance based on, for example, the previous answers they provide to the system. However, creating such content is time-consuming and provides little personal control for the teacher when implemented in an LMS, such as Moodle [21,24]. This ability to adjust feedback to a student's zone of *proximal development* [28] is often considered the gold standard of education that current digital tools and systems, unfortunately, do not meet. Even so, more and more teachers are relying on online activities for instruction. An important feature of this approach is the teacher's ability to not only monitor student progress but also target struggling students for early intervention. Doing this online also allows immediate response and communication between teacher and students when adjusting instructions, moderating difficult learning content, and addressing student misunderstandings [2]. Additionally, an instructor can get an idea of a student's level of engagement in a course by observing the relationship between that student's performance and their use of Moodle [6]. However, the relationship between grades and behavioral data such as Moodle activity logs is complex and influenced by additional factors [6].

3   Case Study Context and the Learning Ecosystem

AAU operates according to a PBL model which assumes that students learn best when applying theory and research-based knowledge to collaborative working strategies aimed at real-world problems. In any one educational program at AAU, each student must enroll in semester study activities corresponding to a total workload of 30 ECTS credits, where a single ECTS is anywhere between

25 to 30 work hours. These 30 ECTS credits typically include a semester project worth 15 ECTS and three courses worth 5 ECTS each. The mandatory study activities at AAU (i.e., semester projects and courses) require students to make connections between them that span from course-to-course in a single semester as well as across semesters.

In the MEA program, the introductory programming course required in the first semester constitutes an important building block for a student's academic success in further semesters. The programming course is offered at two campus locations, one in Aalborg (AAL) and one in Copenhagen (CPH). While the course is primarily held for MEA students, in AAL a group of non-technical Product and Design Psychology students (AAL$_P$) must also enroll as part of their degree program. At both locations, instructors make an effort to "harmonize" their shared syllabus and content, however, their teaching methods may differ.

At AAL, the instructor integrated several methods of flipped instruction [16] in the Fall semester 2017 that were not used in CPH. These methods include online self-study activities consisting of self-assessment quizzes (SA), exercises on khanacademy.org (KA), and mandatory hand-in assignments on peergrade.io (PG). A Moodle course page served as the LMS for providing access to these self-study activities in addition to a collection of other learning resources, such as supplementary video content. The course utilized a combination of online instructions and face-to-face lessons, and when combined with these self-study activities, formed a learning ecosystem which encourages student learning beyond the boundaries of the classroom [7,13,14]. Figure 1 shows an overview of this learning ecosystem used by AAL in its introductory programming course.

This learning ecosystem consisted of several online self-study activities, including course readings, videos, SA and KA, which encourage student learning prior to class. During lectures, the teacher presented the topic in a shortened format, followed by hands-on programming exercises in which the students may

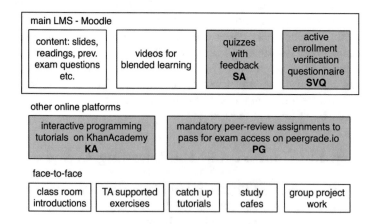

Fig. 1. Overview of the learning ecosystem used by AAL campus location in its introductory programming course in the Fall semester 2017.

either work alone or in groups with the help of the teacher, teaching assistants, or their peers as part of study cafes. These in-class opportunities were designed to reinforce the concepts learned in the self-study activities through practical experience. At various points in the semester, students were asked to complete programming assignments and evaluate those created by their peers in PG. Such peer learning provided students with the opportunity to critically apply their knowledge.

MEA students at AAU have diverse backgrounds (e.g., in nationality, high school specialization, and proficiency in math) and study interests (e.g., in design or programming) [4]. In order to gather information about this diversity in first-year AAL students useful as control variables, we designed a survey called the *Study Verification Questionnaire (SVQ)*. It consisted of a set of 111 self-reported questions based on established factors for student retention discussed in Sect. 2, such as grit and study habits, among others.

4   Data Collection and Method

In order to discover possible assessment predictors for identifying struggling students, we gathered data from all AAL students for each of the self-assessment activities shown in grey in Fig. 1. Additionally, we collected the students' scores from MT and the final exam from both AAL and CPH. In Sect. 5, we explore the use of these assessments, SVQ, SA, KA, PG, and MT, in the construction of a multiple linear regression model for predicting final exam scores.

Students from AAL_M, AAL_P, and CPH_M all took the same MT and the same final exam. However, not all students who completed MT took the final exam. Some students in AAL_M, for example, were ineligible to attend the final exam due to unfulfilled course requirements, e.g., not completing PG hand-in assignments. Following MT, we invited AAL students with the lowest MT scores to attend tutoring sessions as a first step towards providing early and targeted

Table 1. Number of students in the AAL and CPH campus locations and their average scores on the midterm (MT) and final exams.

		AAL_M	AAL_P	CPH_M	Total
MT	Avg. score	66.67	73.72	63.65	65.78
	Failed students	40	4	36	80
	Total students	84	21	82	187
Tutoring	Not invited	36	17	–	53
	Invited	48	4	–	52
	Attended	7	2	–	9
Exam	Avg. score	44.00	47.86	43.96	44.66
	Failed students	31	5	29	65
	Total students	72	22	83	177

academic intervention. Unfortunately, only a handful of students attended one or more of these sessions. Table 1 shows the number of students in AAL (both educations) and CPH who completed the MT and attended tutoring, as well as the average final exam scores for these students.

We began the analysis of our assessment data by investigating how well MT alone could predict the final exam scores of students at both AAL and CPH. From here, we explored additional predictors related to student engagement and learning progress (i.e., SA, KA, and PG) found in the learning ecosystem implemented at AAL. We concluded our analysis by seeing how well self-reported measures in SVQ could improve the prediction of final exam scores at AAL.

5   Results

In order to predict the final exam scores, we began by using the students' scores on MT from both AAL and CPH to construct two base linear regression models. The results from these initial models were rather poor when using adjusted r-squared as the measure of performance (AAL $r^2 = 0.50, p < 0.001$, CPH $r^2 = 0.46, p < 0.001$). However, due to the learning ecosystem implemented by the AAL campus location, we were able to test the significance of including the additional assessments used by AAL and compare their performance to the base models for both AAL and CPH. From the AAL base model, we constructed a multiple linear regression model by selecting from our additional assessments, SA, KA, and PG. The selection of terms in this model was done using a bi-directional step-wise method using AIC as the selection discriminator. During this process, we used the number of SA assignments completed, the number of KA attempts, and the PG score. Although both SA scores (and not number of attempts or completed assignments) and PG scores proved to be significant alone, they did not improve the overall model using our chosen method for model selection. The best model from this selection method was MT + KA ($r^2 = 0.53, p < 0.001$). Figure 2 shows a comparison of the significant predictor KA and the insignificant predictor SA used in our base model for AAL students who did and did not pass the final exam. Figure 3(a) shows the insignificant predictor PG in our base model for AAL students who did and did not pass the final exam.

Before exploring the role of all SVQ questions in our model, we considered that because the focus of the course was programming, it would be worth seeing the effect that self-reported programming experience (SVQ_{pe}) had on the improved model of MT + KA. We began by manually selecting from SVQ the two questions related to student self-reported programming experience and averaged their values. The addition of this predictor to MT + KA resulted in a marginal improvement ($r^2 = 0.55, p < 0.001$). Figure 3(b) shows the significant predictor SVQ_{pe} used in our improved model for AAL students who did and did not pass the final exam.

In our approach, we have elected to manually select individual predictors from SVQ that improve our model of MT + KA + SVQ_{pe} using adjusted r-squared as a selection criteria. We subsequently dropped each predictor after testing so that at any one time, only a single predictor is used. Following this procedure, we arrived at five SVQ predictors that together significantly improved our model (SVQ$_{subset}$). These predictors each belong to one of five different categories related to a student's behavior and psychology. These categories were *personal trait comparison* [15,19], *growth mindset* [11], *high school trust* [8,29], *attitude towards education* [22], and *understanding of the Media Technology education* [4]. Our final model and best fitting linear regression, AAL$_{final}$, consists of the following assessment predictors: MT + KA + SVQ$_{pe}$ + SVQ$_{subset}$ ($r^2 = 0.72, p < 0.001$). For comparison, an ANOVA test revealed a significant difference from our AAL base model shown above and AAL$_{final}$ ($F = 9.32, p < 0.001$).

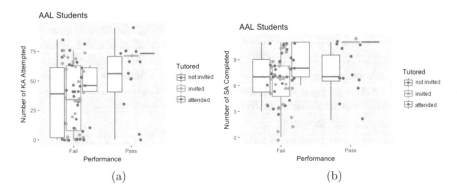

Fig. 2. Box plots grouped according to AAL student performance (pass or fail) on the final exam and tutoring status. Plots show the spread of number of attempted KA assignments in (a) and the number of completed SA assignments in (b).

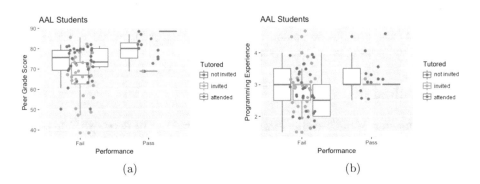

Fig. 3. Box plots grouped according to AAL student performance (pass or fail) on the final exam and tutoring status. Plots show the spread of PG scores in (a) and SVQ$_{pe}$ in (b).

6   Discussion

While the performances of the two base linear models for AAL and CPH were roughly the same, our final AAL model, containing the assessment predictors $MT + KA + SVQ_{pe} + SVQ_{subset}$, proved significantly better than the base CPH model (AAL $r^2 = 0.72, p < 0.001$, CPH $r^2 = 0.46, p < 0.001$). When building our final model, the improvement to the AAL base model by KA and not, for example, by SA, as shown in Fig. 2 could be due to the greater number of possible assignments in the former (103 KA assignments and 42 SA assignments). Even considering their difference in number, students attempted a proportionally greater number of KA assignments than SA.

There are likely a number of reasons for why SA and PG were not selected. With SA, students are allowed to re-take each exercise as many times as they would like. This might encourage "high-score seeking" behavior over actual knowledge retention. There was also a significant drop off in the number of completed SA assignments following the midterm. While PG marginally improved the model, it was not significant given our chosen method for model selection. This could be due to peer assessments being incomparable with one another, whereas having only one reviewer (e.g., a teacher) might ensure a more similar assessment across assignments.

It should be noted that in initial tests using all of SVQ, 40 of the total 111 questions were found to be linearly dependent. This finding of multicollinearity suggests a high degree of redundancy in the questions that could be improved in future uses of SVQ. Additionally, our number of predictor variables in SVQ far exceeds the number of observations. For these reasons, a step-wise selection model may not be wholly appropriate. An alternative method of model selection such as Lasso or Principal Component Analysis (PCA) would likely be more robust to such situations.

Overall, our results indicate that while student performance on MT was positively correlated with performance on the final exam in the AAL and CPH base models, having a learning ecosystem which consists of several appropriate and diverse assessments, as demonstrated by the final AAL model, significantly improved the prediction of final exam scores.

7   Conclusion

The instruments of the learning ecosystem presented in this paper provide initial findings in support of additional strategies for targeting struggling students in a PBL environment. While the results leave much room for improvement, they nonetheless demonstrate that regular student feedback through self-regulated knowledge assessments, targeted tutoring, and proper evaluations of student behavior and psychology may be essential factors in reducing rates of PBL student failure. Moreover, technological learning tools through, e.g., Moodle, Peergrade or Khan Academy, might serve as useful tools for ensuring the academic success of students. Such benefits are particularly needed at universities where more and more degree programs are becoming interdisciplinary and courses are being taught by different instructors at separate campus locations. Guaranteeing the quality of education in these situations is essential.

8   Future Work

In the future, we hope to implement the significant assessments of our learning ecosystem into a system for identifying struggling students prior to the midterm of a given course and incorporate additional sources of relevant information such as Moodle course activity. The variety of significant assessment predictors in our final model emphasizes the need for a learning ecosystem that is both targeted and wide-ranging, as shown, for example, in Fig. 1. With this model, it might be possible in future semesters to target struggling students even before the start of a course by identifying those who reported a low SVQ_{pe} score. In compliance with LTI, an automated weekly analysis could invite students in need to group tutoring sessions based on how often students either attempted KA or completed SA assignments.

References

1. Banas, J. R., Velez-Solic, A.: Designing effective online instructor training and professional development. In: Virtual Mentoring for Teachers: Online Professional Development Practices: Online Professional Development Practices, p. 1 (2012)
2. Bergmann, J., Sams, A.: Flipping for mastery. Educ. Leadersh. **71**, 24–29 (2014)
3. Bloom, B.S., Engelhart, M.D., Furst, E.J., Hill, W.H., Krathwohl, D.R.: Taxonomy of educational objectives: the classification of educational goals. In: Handbook I: Cognitive Domain. David McKay Company, New York (1956)
4. Bøgelund, P., Justesen, K., Kolmos, A., Bylov, S.M.: Undersøgelse af frafald og fastholdelse ved medialogi og andre uddannelser ved det Teknisk-Naturvidenskabelige Fakultet 2015–2016: Arbejdsrapport Nr. 1. Aalborg Universitet (2016)
5. Cabrera, A.F., Nora, A., Castaneda, M.B.: College persistence: structural equations modeling test of an integrated model of student retention. J. High. Educ. **64**(2), 123–139 (1993)
6. Casey, K., Gibson, P., Paris, I.: Mining moodle to understand student behaviour. In: International Conference on Engaging Pedagogy (2010)
7. Chang, M., Li, Y.: Smart Learning Environments. Lecture Notes in Educational Technology. Springer, Heidelberg (2015)
8. Cohen, G.L., Garcia, J., Purdie-Vaughns, V., Apfel, N., Brzustoski, P.: Recursive processes in self-affirmation: intervening to close the minority achievement gap. Science **324**(5925), 400–403 (2009)
9. Dekker, G., Pechenizkiy, M., Vleeshouwers, J.: Predicting students drop out: a case study. In: Proceedings of Educational Data Mining (2009)
10. Duckworth, A.L., Quinn, P.D.: Development and validation of the short grit scale (Grit-S). J. Pers. Assess. **91**(2), 166–174 (2009)
11. Dweck, C.S.: Self-theories: Their Role in Motivation, Personality, and Development. Psychology Press, Philadelphia (2000)

12. Evans, C., Palacios, L.: Interactive self assessment questions within a virtual environment. Int. J. e-Adoption (IJEA) **3**, 1–10 (2011)
13. Giovannella, C.: Smart learning eco-systems: "fashion" or "beef"? J. e-Learn. Knowl. Soc. **10**(3), 15 (2014)
14. Giovannella, C., Rehm, M.: A critical approach to ICT to support participatory development of people centered smart learning ecosystems and territories. Aarhus Ser. Hum. Centered Comput. **1**(1), 2 (2015)
15. Glynn, J.G., Sauer, P.L., Miller, T.E.: A logistic regression model for the enhancement of student retention: the identification of at-risk freshmen. Int. Bus. Econ. Res. J. **1**(8), 79–86 (2011)
16. Green, L.S., Banas, J., Perkins, R.: The Flipped College Classroom: Conceptualized and Re-Conceptualized. Springer, Heidelberg (2016)
17. Guerrero, W.: Flipped classroom and problem-based learning in higher education. In: Latin-American Context, Conference Proceedings. The Future of Education, p. 118 (2017)
18. Herzog, S.: Measuring determinants of student return vs. dropout/stopout vs. transfer: a first-to-second year analysis of new freshmen. In: Proceedings of 44th Annual Forum of the Association for Institutional Research (AIR) (2004)
19. Higher Education Research Institut: CIRP Freshman Survey – HERI (2017). https://heri.ucla.edu/cirp-freshman-survey/
20. Lassibille, G., Gomez, L.N.: Why do higher education students drop out? Evidence from Spain. Educ. Econ. **16**(1), 89–105 (2007)
21. Leal, J.P., Queirós, R.: Using the learning tools interoperability framework for LMS integration in service oriented architectures. In: Technology Enhanced Learning, TECH-EDUCATION 2011 (2011)
22. Levitz, R.N.: Retention management system plus samples survey and report samples (2012). https://www.ruffalonl.com/complete-enrollment-management/student-success/rnl-retention-management-system-plus/samples
23. Mayer, R.E.: Principles for managing essential processing in multimedia learning: segmenting, pretraining, and modality principles. In: The Cambridge Handbook of Multimedia Learning, pp. 169–182 (2005)
24. Severance, C., Hardin, J., Whyte, A.: The coming functionality mash-up in personal learning environments. Interact. Learn. Environ. **16**(1), 47–62 (2008)
25. Seymour, E., Hewitt, N.: Talking About Leaving. Westview Press, Boulder (1997)
26. Touron, J.: The determination of factors related to academic achievement in the university: implications for the selection and counseling of students. High. Educ. **12**, 399–410 (1983)
27. Tsukayama, E., Duckworth, A.L., Kim, B.: Domain-specific impulsivity in school-age children. Dev. Sci. **16**(6), 879–893 (2013)
28. Vygotsky, L.: Zone of proximal development. In: Mind in Society: The Development of Higher Psychological Processes, p. 157 (1987)
29. Yeager, D.S., Purdie-Vaughns, V., Garcia, J., Apfel, N., Brzustoski, P., Master, A., Hessert, W.T., Williams, M.E., Cohen, G.L.: Breaking the cycle of mistrust: wise interventions to provide critical feedback across the racial divide. J. Exp. Psychol. Gen. **143**(2), 804–824 (2014)

Automated Prediction of Student Participation in Collaborative Dialogs Using Time Series Analyses

Iulia Pasov[1,2], Mihai Dascalu[1,3(✉)], Nicolae Nistor[2,4],
and Stefan Trausan-Matu[1,3]

[1] Faculty of Automatic Control and Computer Science,
University "Politehnica" Bucharest,
313 Splaiul Independenței, 60042 Bucharest, Romania
iulia.pasov@cti.pub.ro, mihai.dascalu@cs.pub.ro,
stefan.trausan@cs.pub.ro
[2] Faculty of Psychology and Educational Sciences,
Ludwig-Maximilians-Universität, Leopoldstr. 13, 80802 Munich, Germany
nic.nistor@uni-muenchen.de
[3] Academy of Romanian Scientists,
Splaiul Independentei 54, 050094 Bucharest, Romania
[4] Richard W. Riley College of Education and Leadership, Walden University,
100 Washington Avenue South, Suite 900, Minneapolis, MN 55401, USA

Abstract. The massive student participation in Computer Supported Collaborative Learning (CSCL) sessions from online classrooms requires intense tutor engagement to track and evaluate individual student participation. In this study, we investigate how the time evolution of messages predicts students' participation using two models – a linear regression and a Random Forest model. A corpus of 10 chats involving 47 students was scored by 4 human experts and used to evaluate our models. Our analysis shows that students' pauses length between consecutive messages within a discussion is the strongest participation predictor accounting for $R^2 = .796$ variance in the human estimations while using a Random Forest model. Our results provide an extended basis for the automated assessment of student participation in collaborative online discussions.

Keywords: CSCL · Time series analysis
Automated evaluation of participation

1 Introduction

The concept of smart environment has been used in many different domains, from homes, vehicles, to cities or education. According to Murillo Montes de Oca et al. [1], a smart learning ecosystem is defined not only to be technologically enhanced, but also focused on participants and collaboration. In such ecosystems, the interconnection between people and technology is meant to contribute to the success of all participants throughout the learning process. This is achieved by using Computer Supported

© Springer International Publishing AG, part of Springer Nature 2019
H. Knoche et al. (Eds.): SLERD 2018, SIST 95, pp. 177–185, 2019.
https://doi.org/10.1007/978-3-319-92022-1_16

Collaborative Learning (CSCL) technologies. As students' lives are constantly influenced by technology, especially in regards of socialization, they expect the same transition in education. They often prefer synchronous activities because they have a better experience of active involvement and social connection. Tutors also use chats as assignments for tasks which do not have a straightforward solution, and require brainstorming or discussions for choosing between possible solutions. In such situations, it is very important that students are engaged in discussions and participate to building a solution.

Earlier studies by Ferschke et al. [2, 3] identified both positive and negative effects when using chats within online courses. Most negative effects occur due to time mismatches, as not all students can participate synchronous at a moment in time. The studies raised questions on how to exploit the benefits of synchronous social interaction while avoiding coordination difficulties.

Important limitations of technology enhanced classrooms are related to the human side of the educational experience because teachers have limited time to manually evaluate students' participation [4]. Moreover, training more tutors to support, monitor and properly evaluate students' participation to discussions takes time and it is not always feasible. Indicators like performance, collaboration or participation with other members need to be constantly monitored to ensure each student's benefit of the collaborative learning process. In classical learning environments, such evaluation is time consuming and its complexity increases with the number and variety of students registered for the activity, but smart environments are meant to provide automated tools designed to estimate student's collaboration and participation, at any scale, in real time, allowing tutors to automatically monitor and assess the collaborative learning process.

Sfard [5] defines learning by two metaphors: acquisition and participation. While the first one relates to learning as an individual knowledge acquiring process, the participation metaphor states that learning consists of increasing participation in dialog construction, for example, in communities of practice. Therefore, learning can be done either independently or as a social knowledge building process [6, 7]. Collaborative learning happens when students participate to discussions, and just like in the case of the communities of practice, participation leads to accumulation of experience and stimulates the social construction of knowledge [8]. This makes student participation an important metric that needs to be observed by tutors during collaborative assignments. Participation in CSCL contexts has been extensively studied by Dascalu et al. [9] using the Cohesive Network Analysis model which considers an in-depth perspective of dialogs.

The focus of this article is to predict student participation in CSCL chats using features that account for discussion changes in time. Although prediction of student participation in chats is not a new topic [3, 4, 9, 10], few studies focused on features related to time. Boroujeni et al. [11] used time series to analyze the impact of regularity on academic success and emphasized the importance of time features such as the day of the week or month, to quantify regularity and predict performance. However, in the case of CSCL chats (1–2 h discussions) such features could only be used if the corpus displays enough variance such that no false conclusions are drawn. Therefore, we need to establish a representation for features derived from time series which can be used in automatic assessment of individual participation.

2   Research Question

In this study, our aim is to answer the following question: Which are the most relevant time-related features in automatic assessment of student participation in chat conversations? Are these global features derived from the entire chat discussion, or specific to each participant?

Previous research by Molomer et al. [10] shows the importance of pauses in predicting student participation. Building on these findings, we investigate whether the pause length between utterances is positively or negatively related to participation. A positive relationship could be explained by longer reflection which, in turn, can strengthen participation, whereas a negative relationship may be due to fast replies which would be associated with increased participation. The latter would be a trivial result.

3   Method

3.1   Conversation Corpus

The corpus used for this study contains ten chat discussions selected from a larger corpus of over 100 undergraduate students' transcripts. All participants were students in the same year, which previously knew each other. Four to five students had to debate over the importance of several CSCL technologies: chat, forum, blog, wiki, wave (Google Wave). Each student had to convince the others of the advantages of his/her chosen technology by also presenting the disadvantages of other technologies. In the second part of the discussion, the participants searched for an integrated technical solution that includes most of the discussed advantages from the first part [4].

Student participation or active involvement was scored afterwards by four human experts with values from 1 to 10 (1 – almost no involvement; 10 – overall great participation). The average intraclass correlation score (ICC) between raters was 0.75, with 95% confidence for the interval 0.60–0.85. This proves that there is agreement between raters while relating to the participation scores. The participation score used further for this study is the average value between experts' individual participation values.

3.2   Data Analysis Methods

Feature extraction. We focused on two types of features: those extracted from the chat and those derived from participants' contributions. We tried to identify the most relevant features from each chat which can be used in predicting participation, starting with the ones used by Molomer et al. [10]. Only three features were used in the regression models to avoid overfitting, given their relationship with the participation score and between each other.

The chat-related features extracted from the discussion were: (a) Participants – the number of participants in the chat, (b) Time of the day – the time of the day when the conversation started, (c) Duration – the total duration in minutes, and (d) Frequency – the average distance between consecutive utterances in the chat, measured in seconds. For each property, Spearman's rank correlations with corresponding significance levels were used to test the correlations between the average participation score and the extracted feature.

Table 1. Spearman's rank correlations and p-value between chat-related features and student participation.

Property	ρ	p
Time of the day	.17	.23
Duration	.12	.39
Frequency	.07	.59
Participants	−.27	.06

Several features were extracted from each participant's time series of utterances. Since the chats were requested as an assignment, in a proposed time interval, general time-based information about the year, month or day of the week was not relevant. Moreover, the discussions were relatively short, so the participants could schedule them according to their convenience, and we could not propose any seasonality-based features. For each participant, we considered: (a) Contributions – the number of contributions in the chat, (b) Pauses – time series of pauses between consecutive utterances, measured in seconds, (c) Out – time series created from the number of referred utterances for each message, (d) In – time series created from the number of utterances which referred each utterance in the chat and (e) Words – time series with the count of words in utterances, without stop-words.

In classical time series analysis, Brockwell et al. [12] broke time series into trends and seasonal patterns. When time seasonality is difficult to identify, only the trend can be estimated. Therefore, only the series' linear trend was considered as a relevant feature in our time series analysis. Trends were estimated using linear regressions and from their equations, the following parameters were extracted: coefficients and intercepts, thus obtaining two new features. A separate feature (coefficient positive) has been introduced because the sign of the coefficient brings extra information: a negative coefficient means that with the increase of the feature's value, participation decreases, while positive values reflect similar trends. The sum and average of all the time series (b–e) are presented as independent features, as well.

Spearman's rank correlation coefficient between the participant-related features and student participations are provided in Table 2. From the correlation analysis, it is revealed that all participation-related features impact, at some level, students' participation. While relating to the pause between utterances, both intercept and average values appear to be strongly correlated to participation. Moreover, a negative coefficient – an overall decreasing pause in time – between utterances, seems to increase student participation. However, we must emphasize a limitation of our study - the number of contributions is significantly related with the participation score.

The feature most correlated to student participation is the average pause according to Spearman's rank correlation coefficient from Table 2, followed by the number of contributions and the total number of words a participant contributes with.

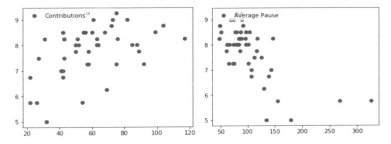

Fig. 1. Visual correlation between the number of contributions and average pause (x-axis), and average participation (y-axis).

Figure 1 shows that participants who reply faster, with pauses of 40–70 s, are perceived to participate more within the conversation. Participants with longer average pauses between utterances (over 2 min) seem to participate less (score less than 8.5). Thus, the more students contribute, the higher their participation score is.

The following features with the highest Spearman's rank were tested against each other for cross-correlations in Table 3: average pause, pause linear intercept, sign of the pause linear coefficient, total number of referred utterances, total number of words

Table 2. Spearman's rank correlation and p-value between participant related features and student participation.

Property	ρ	p-value
Pause Coefficient	−.17	.24
Pause Coefficient positive	−.25	.08
Pause Intercept	−.41	<.01**
Average Pause	−.51	<.01**
Total Pause	−.08	.59
Out Coefficient	−.13	.38
Out Coefficient positive	−.10	.48
Out Intercept	−.04	.76
Average Out	.02	.89
Total Out	.32	.02**
In Coefficient	−.06	.68
In Coefficient Positive	−.04	.75
In Intercept	−.14	.32
Average In	.22	.13
Total In	−.18	.22
Words Coefficient	.04	.77
Words Coefficient Positive	.06	.66
Words Intercept	−.22	.13
Average Words	−.00	.99
Total Words	.40	<.01**
Contributions	.46	<.01**

submitted by the participant, and the contributions (utterances). Features strongly correlated to participation also exhibit a high correlation among themselves ($\rho > .6$).

Table 3. Spearman correlations between top features.

Property	AvgP	PI	PC > 0	TotOut	TotW	TotC
Average Pause (AvgP)	1	.61	−.21	−.65	−.71	−.81
Pause Intercept (PI)	.61	1	−.65	−.33	−.56	−51
Pause Coefficient positive (PC > 0)	.21	−.4	1	−.26	−.05	−.08
Total Out (TotOut)	−.65	−.33	−.26	1	.77	.71
Total Words (TotW)	−.71	−.56	−.05	.77	1	.83
Contributions (TotC)	−.81	−.51	−.08	.71	.83	1

4   Results

We propose two models to predict students' participation score: linear regression and random forests (RF) [13]. Linear regression has been used before by Dascalu et al. [9] to predict participation, with $R^2 = .433$ ($p < .001$), whereas Molomer et al. [10] reported $R^2 = .546$ ($p < .001$) using a similar dataset.

For the linear regression, the first three most relevant features, not correlated one with another, were chosen. The features most correlated to student participation according to Tables 1 and 2 are: (a) the average pause, (b) the number of contributions, and (c) the total number of words. However, according to Table 3, all three are highly correlated with corresponding similarity scores between .71 and .83. A stepwise linear regression was performed using the backward method and starting with all the proposed features from Tables 1 and 2, until the selection was limited to: (a) the average pause, (b) the intercept for the time series of words, and (c) the number of participants. With this approach, we obtained a significant model $R^2 = .502$ ($p < .001$). With a leave-one-out cross-validation, the linear regression produced a model explaining around 47% variance ($R^2 = .473$ $p < .001$).

We decided to analyze also a second model for participation estimation, one that can use non-linear dependencies between features, namely a Random Forests (RF) model [13]. The first step consists of training the RF model with all the features to establish their importance. With this new approach, pauses between students' utterances have the highest importance score for predicting participation (39%), followed by the number of words, the number of utterances, the linear intercept in pauses and the frequency. Afterwards, using the same 3 features as in the case of the linear regression, a maximum depth of 4 and 2 estimators, our model explained around 77% variance ($R^2 = .769$, $p < .001$) or, in the case of leave-one-out cross-validation, $R^2 = .499$ ($p < .001$). Compared to the average participation scores, the mean average error for the prediction is .39, thus denoting close scores to human expert judgements. A similar RF model trained on frequency, average pause and words intercept explained $R^2 = .796$ ($p < .001$) variance, while the leave-one-out cross-validation had $R^2 = .643$ ($p < .001$).

5   Discussion and Conclusion

In response to the research question, a limited number of features, independent one from another, were used for the time series of less than 100 data points in order to avoid overfitting. The regression models provided accurate estimations, similar to the scores provided by the experts. Correlations were found between chat-related and participant-related features on one hand, and the student participation score, on the other hand (see Tables 1 and 2). One of the features which stood out during the analysis is the average pause between student's consecutive utterances. Longer pauses can influence students' participation in a negative way, when the student takes too long to reply to messages, or a positive way, when the student's pauses are long due to his mental preparation for high quality responses. In our analysis, we found a negative correlation score between the average pause and participation, which supports the first statement. Therefore, students who reply faster and more often are more likely to have a higher participation into the chat discussion.

The next two features, highly correlated with student participation are the number of contributions and the number of words, both with positive coefficients, which means that participation increases with a higher number and longer messages. Both features prove that volume, not only speed, have a high impact on student participation. However, both are highly correlated with the average pause: students who reply faster submit more messages and therefore more content (words). The high correlation between those properties explains why, during the stepwise analysis, the best choice for the linear regression was a different selection.

Since the top five features most correlated to student participation are all participant related, according to Spearman's correlation coefficients computed in Tables 1 and 2, this proves that participation is more related to the student's input, and the impact of others is not as high as their own social contribution. The only chat-related feature with a high correlation score with student participation is the number of participants in the chat. The course assignment imposed between 4 to 5 students. The negative ρ coefficient proves that the participation score decreases with a higher number of participants. Moreover, there are no members with low participation scores (5 to 7.5) in chats with fewer participants. This can also be observed in the correlation scores from Table 1. However, studies in cooperative learning [14, 15] show that there might be other reasons: the distribution of gender in the chat group, previous cooperation of the participants, previous knowledge about the topic, communication skills or the quality of peer interaction.

With the help of the proposed properties, regression models were trained to predict student participation. The selection for the linear regression generated similar results to previous studies by using the following features: the average pause, the intercept for the time series of word counts, and the number of participants. An improvement to the study by Molomer et al. [10] comes from the different regression model and properties selection used to predict participation scores, as seen in Table 4.

The random forests model estimated more precisely student participation. As such methods can find non-linear correlations, the result only argues that there are more complex relations between the chosen features. Both the number of participants and the

Table 4. Regression results for participation score.

Method	Features	R^2
Linear regression	Pauses Total number of words Fluency	.546
Linear regression	Average pause Intercept for the number of words The number of participants	.502
Random forest	Average pause Intercept for the number of words The number of participants	.769
Random forest	Frequency Average pause Intercept for the number of words	.796

frequency from the chats have low correlation ranks with participation, but combined with the average pause and the intercept of the time series of word count, they provide more accurate models. Since one out of three properties in the proposed regression models is a chat-related property, this explains that participation is not only influenced by students taken individually, but also by the environment they are part of.

One potential next step consists of making our models easily available to tutors who are interested in monitoring student participation. According to Sfard [5], participation has the potential to improve the practice of learning and of teaching. The proposed models based only on features which can be easily computed at any moment in time (e.g., average values or linear intercepts) greatly facilitate real time feedback generation to stimulate greater participation within the dialog. This also facilitates the generation of tutors or automated interventions meant to guide students in their collaborative learning tasks.

Acknowledgements. This research was partially supported by the 644187 EC2020 Realising an applied Gaming Eco-system (RAGE) project.

References

1. Murillo Montes de Oca, A., Nistor, N., Dascalu, M., Trausan-Matu, S.: Designing smart knowledge building communities. Int. J. Interact. Des. Archit. **22**, 9–21 (2014)
2. Ferschke, O., Howley, I., Tomar, G., Yang, D., Rosé, C.P.: Fostering discussion across communication media in massive open online courses. In: Proceedings of Computer Supported Collaborative Learning, pp. 459–466 (2015)
3. Ferschke, O., Yang, D., Tomar, D., Penstein Rose, C.: Positive impact of collaborative chat participation in an edX MOOC. In: International Conference on Artificial Intelligence in Education, pp. 115–124. Springer, Madrid (2015)
4. Trausan-Matu, D., Dascalu, M., Rebedea, T.: PolyCAFe—automatic support for the polyphonic analysis of CSCL chats. Int. J. Comput. Support. Collaborative Learn. **9**(2), 127–156 (2014)

5. Sfard, A.: On two metaphors for learning and the dangers of choosing just one. Educ. Res. **27**(2), 4–13 (1998)
6. Stahl, G.: Group Cognition: Computer Support for Building Collaborative Knowledge. MIT Press, Cambridge (2006)
7. Trausan-Matu, S., Stahl, G., Sarmiento, J.: Supporting polyphonic collaborative learning. E-Serv. J. **6**(1), 58–74 (2007)
8. Nistor, N., Baltes, B., Dascalu, M., Mihaila, D., Trausan-Matu, S.: Participation in virtual academic communities of practice under the influence of technology acceptance and community factors, a learning analytics application. Comput. Hum. Behav. **34**, 339–344 (2013). https://doi.org/10.1016/j.chb.2013.10.051
9. Dascalu, M., McNamara, D.S., Trausan-Matu, S., Allen, L.K.: Cohesion network analysis of CSCL participation. Behav. Res. Methods, 1–16 (2017)
10. Molomer, S.D., Dascalu, M., Trausan-Matu, S.: Predicting collaboration based on students' pauses in online CSCL conversations. U.P.B. Sci. Bull. Ser. C **79**(2) (2017)
11. Boroujeni, M. S., Sharma, K., Kidzinski, L., Lucignano, L., Dillenbourg, P.: How to quantify student's regularity? In: European Conference on Technology Enhanced Learning, pp. 277–291. Springer (2016)
12. Brockwell, P.J., Davis, R.A.: Time Series: Theory and Methods. Springer Science & Business Media, New York (2013)
13. Liaw, A., Wiener, M.: Classification and regression by randomForest. R News **2**(3), 18–22 (2002)
14. Webb, N.M.: Sex differences in interaction and achievement in co-operative small groups. J. Educ. Psychol. **75**, 33–44 (1984)
15. Terwel, J., Gillies, R.M., Van den Eeden, P., Hoek, D.: Co-operative learning processes of students: a longitudinal multilevel perspective. Br. J. Educ. Psychol. **71**, 619–645 (2001)

Author Index

© Springer International Publishing AG, part of Springer Nature 2019
H. Knoche et al. (Eds.): SLERD 2018, SIST 95, pp. 187–188, 2019.
https://doi.org/10.1007/978-3-319-92022-1

Printed in the United States
By Bookmasters